THE NUCLEAR WASTE PRIMER

THE LEAGUE OF WOMAN VOTERS
EDUCATION FUND

NUCLEAR WASTE PRIMER

The League of Women Voters Education Fund

For further information, please call:
(800) 225-NWPA
(202) 488-5513 (in the Washington, DC area)

or write:
OCRWM Informtion Center
P.O. Box 44375
Washington, DC 20026

Additional information regarding the U.S. Department of Energy
Waste Isolation Pilot Plant can be obtained by contacting:
The Office of Public Affairs
P.O. Box 3090
Carlsbad, NM 88221
Telephone (506) 885-7337

Printed in the United States of America

Library of Congress Cataloging-in-Publication Data

The Nuclear Waste Primer/the League of Women Voters Education Fund.
p. cm.
Previous ed. published: New York, NY: N.Lyons books, ©1985
Includes bibliographical references and index.
ISBN 1-55821-226-4
1. Radioactive waste disposal-United States.
I. League of Women Voters (U.S.). Education Fund.

TD898.118.N85 1993 93-11363
363.72'89-dc20 CIP

Printed with soy ink on recycled paper

CONTENTS

ACKNOWLEDGMENTS

The Nuclear Waste Primer was written by Susan D. Wiltshire, JK Research Associates, Inc., with assistance from associate Carol Williams and with research assistance provided by Joe Oosterhout and Karen Goxem of the League of Women Voters Education Fund; edited by Monica Sullivan. The project was directed by Elizabeth Kraft. Earlier editions were written by Marjorie Beane and Isabelle Weber. The LWVEF wishes to thank the project advisory committee and the many reviewers of the draft manuscript.

LWVEF Chair: Becky Cain

Project Trustee: Nancy Pearson

Executive Director: Gracia Hillman

Director: Sherry Rockey

Natural Resources Program Manager: Elizabeth Kraft

Publications Director: Monica Sullivan

The Nuclear Waste Primer, part of an LWVEF project on nuclear waste education, was made possible by a cooperative agreement with the U.S. Department of Energy.

HOW TO USE THIS PRIMER

This *Primer* provides information about nuclear waste in the United States—what it is, where it comes from, how it has been managed, and what we can do with it in the future.

This book is written for anyone interested in radioactive waste and its effects. The topic is *not* for experts only. Readers new to the subject may want to read the *Primer* straight through, occasionally flipping to definitions in Chapter 2 or to the glossary. Others will use the *Primer* as a reference for specific topics. The overall organization of the book reflects two broad areas of interest: civilian waste and defense waste. Although their histories intertwine, the management of these two categories of waste has been almost completely separate, and most readers will find themselves concerned primarily with one or the other. Thus, the *Primer* has been organized to make it easy for readers to find the topics that particularly interest them.

The first chapters provide a basic introduction to the issues and technical terms relevant to both civilian and defense waste. Chapter 1 sketches the history of nuclear waste management and lists significant technical and social issues. Chapter 2 discusses radiation and the risks it poses, describes the different categories of waste, and explains how responsibility for nuclear waste management and regulation is divided among government agencies and other entities.

Chapters 3 and 4 concern civilian nuclear waste, which is all nuclear waste generated by nongovernment activities, including using nuclear reactors to produce electricity. Chapter 5 considers nuclear waste transportation and liability for accidents. Chapter 6 concerns defense nuclear waste, which is all nuclear waste generated by the federal government's nuclear weapons program. It explains the cleanup of contaminated sites and the management (treatment, storage, and disposal) of the waste itself. Chapter 7 suggests ways citizens can have an impact on national, regional, and local radioactive waste management policies and programs.

The *Primer* includes a glossary of terms, a list of publications, and a list of resources—organizations concerned with radioactive waste management. The list of publications includes sources that have been used in writing this book and sources for further information. The list of organizations is intended to help readers make the contacts they need to stay informed and to become involved in issues that concern them.

ONE

INTRODUCTION

Nuclear wastes are the radioactive by-products of nuclear weapons production, nuclear power generation, and other uses of nuclear materials. They range from highly radioactive discarded nuclear fuel to slightly radioactive used clothing.

The management of nuclear waste received little attention from government policymakers during the three decades after the atomic bomb's development in 1945. The nation spent billions of dollars to produce nuclear weapons and to commercialize nuclear power in the 1950s and 1960s, while spending only a few hundred million dollars to research storage and disposal processes. However, since the mid-1970s, considerable public and government attention and resources have focused on nuclear waste as a serious national problem.

DEFENSE WASTE

The U.S. government developed the atomic bomb during World War II under conditions of urgency and secrecy. The United

States continued to develop and produce advanced atomic weapons during the subsequent Cold War period under similar conditions of secrecy and with a sense of mission and urgency. In many instances, radioactive wastes were treated, stored, or disposed of with an eye toward expedient short-term measures that created significant waste disposal and site cleanup problems.

Weapons production facilities—including factories, laboratories, and waste disposal sites—were exempt from most outside regulation and largely remained closed to public view. However, the news broke in the early 1970s that liquid high-level radioactive waste was leaking from government storage tanks and that abandoned uranium mill tailings had been dispersed in the environment. This information raised public awareness and concern about how the federal government managed its facilities. Some people, especially neighbors of nuclear facilities, became increasingly concerned about the adequacy of plans for future disposal and the potential environmental and public health impacts of disposal practices.

Successful lawsuits, new legislation, and changed policies gradually opened the weapons production facilities to state and public review and oversight. In the late 1970s, the federal government began allocating substantial funds and personnel to develop a plan for the long-term management of defense waste. In 1987, the Department of Energy (DOE) was reorganized to consolidate defense-related waste management programs and place increased emphasis on site cleanup.

The end of the Cold War in the early 1990s has brought major changes at weapons production sites. In January 1992, President Bush canceled the only nuclear warheads still in production, and Bush's Secretary of Energy, James Watkins, made cleaning up former weapons production facilities and complying with environmental laws a top priority for DOE. Major legislation also has changed the way defense waste management decisions are made, opening up facility operation to outside regulation and the decision-making process to outside participation and review. However, distrust of the federal government's commitment and ability to develop and carry out a safe waste management system continues to complicate relationships among the Department of Energy, regulatory agencies, other state, local, and tribal officials, and the public.

CIVILIAN WASTE

The nuclear power industry developed from the research conducted to produce nuclear weapons. As utilities began to build and operate commercial nuclear power plants in the late 1950s, most scientists, and the regulators and proponents of nuclear power, viewed high-level waste management as a technical problem for which future technology would provide a solution.

Radioactive materials were increasingly used, not only in nuclear weapons production and electric power generation, but also in industrial processes, in medical and biotechnological research, and in diagnosing and treating disease, and many other ways. Users of radioactive material relied on the private sector to provide waste disposal facilities for the resultant low-level radioactive waste.

However, concern rose during the 1970s about possible shortages of storage and disposal capacity for nuclear waste and about the need for a complete, reliable waste management system for both high-level and low-level waste.

In the late 1970s, power plant operators realized that serious storage problems for their reactors' spent fuel could emerge by the late 1980s and that some reactors might have to shut down in the mid-1990s unless additional storage became available either on- or off-site. At the same time, the public began questioning whether the methods and materials used for containing radioactive waste could endure over the very long time periods required for the decay of high-level nuclear waste, including spent fuel. In the 1970s, several states passed laws prohibiting further nuclear power plant construction until the federal government had demonstrated that waste could be disposed of safely and permanently. Other states restricted or prohibited disposal of radioactive waste within their borders.

As for low-level waste, by 1978 only three operating commercial disposal sites in the United States—at Barnwell, South Carolina; Beatty, Nevada; and Richland, Washington—remained to receive low-level radioactive waste from all of the country's nuclear power plants, medical and research facilities, and other industries. Governors of those states gave notice that they planned either to close their facilities or cut back on the amount of low-level waste the sites would accept in the future.

In this atmosphere of concern, Congress began considering comprehensive radioactive waste legislation in the late 1970s. Several sessions of Congress wrestled with a series of tough issues, gradually developing policy options that provided the basis of compromise legislation: the Low-Level Radioactive Waste Policy Act of 1980 and the Nuclear Waste Policy Act of 1982 (discussed in Chapters 3 and 4). Together these laws provided an initial framework for resolving radioactive waste management issues.

Within five years of passage, each law was substantially amended as citizens and public officials continued to struggle with contentious nuclear waste problems. The Low-Level Radioactive Waste Policy Amendments Act of 1985 gave states more time to develop the disposal facilities required by the 1980 act, but it also imposed a strict timetable and heavy penalties for missing milestones. The Nuclear Waste Policy Amendments Act of 1987 directed DOE to investigate only one potential repository site for high-level nuclear waste at Yucca Mountain, Nevada, and established a separate federal agency to seek volunteer states and Indian tribes willing to host a temporary spent fuel storage or disposal site. The Nuclear Waste Policy Act also specified roles for state, tribal, and local governments to play in making the complex policy decisions about civilian radioactive waste management.

By the end of 1992, several major milestones set in the 1985 Low-Level Radioactive Waste Policy Amendments Act and the 1987 Nuclear Waste Policy Amendments Act had been missed or were likely to be missed. Thus, changes or additions to current policy and legislation seem probable.

ISSUES

Resolving issues and forging public policy on radioactive waste is difficult for many reasons. The responsibility for setting policy is widely dispersed among federal and state governments and agencies. The debate about nuclear waste raises complex social issues, such as regional and generational equity. And although much that is known about radioactive waste can be called scientific fact, much else is far less certain. For example, some predictions—such as those about the effects of low levels of radiation on human health and about the long-term behavior of geologic and manufactured systems—necessarily incorporate uncertainty, a difficult context for policy discussions.

Nuclear waste management raises numerous basic questions. Do the benefits of an activity justify producing waste? If so, how can the amount of waste be minimized? How and where should the waste be stored? Should the waste be treated in some way to make it safer to handle, store, and dispose of? How and where should it be treated? When and why should waste be moved? How and by what routes should waste be transported? Can the waste be successfully isolated? Where and how?

Answering these questions requires a set of measures against which to judge alternative answers. How hazardous is the waste? How much exposure (what level of risk) should be allowed to workers, the public, or the environment? How reliable are various methods of management, disposal, or transportation? Is the cost in dollars and in risk worth the benefit? How should the risk of doing nothing be balanced against the risk of available options?

Whenever a new facility must be sited to provide treatment, storage, or disposal of nuclear waste, more questions arise. If such facilities are necessary, how should sites for them be found? Through a purely technical process, a political process, a volunteer process? How can a siting process be made fair to potential host communities? Should we look for the "best" site (a term that is impossible to define) or for an acceptable and adequate site? What is the "best" treatment, storage, or disposal method? Why should a state, tribe, or locality accept any risks posed by a waste facility when the activities that generated this waste benefitted people elsewhere? How can the risks be reduced and a community's rights protected?

Cleaning up contaminated sites raises other issues. What should be the goal of clean-up? What eventual uses of contaminated land are achievable or acceptable? How clean can we expect to get a site? How much money can or should we spend to achieve that result? Should the contaminated material that results from a cleanup be moved from the site? Where should it go?

Behind all these questions are the basic issues of moral obligation, responsibility, and trust. Is the current generation morally obligated to proceed with permanent waste disposal in the face of uncertainty, or should we delay decisions in the hope of developing better technology, knowing that delay burdens future generations with responsibility for disposal? How should we compare risks to the health of today's workers and to the public with risks to the health of future generations? Who can vouch for the safety of first-of-a-kind facilities that must isolate the waste for hundreds or hundreds of thousands of years?

No one can be expert in all the fields of science and technology that come into play in radioactive waste management decisions. We all must rely on others for some information and judgments. Whom can we trust to provide good answers? If most scientists currently agree on an issue, but counterarguments are strong, how can we tell who is right? How can we make decisions that incorporate both scientific fact and social values?

One purpose of this book is to provide the reader with information for thought and discussion about these issues. Sometimes political or value considerations are hard to separate from scientific ones. Both are important elements of waste management decisions. Politically acceptable decisions that are scientifically unsound will not serve the public interest; nor will scientifically valid decisions suffice if they are incompatible with public values. Our challenge as citizens is to do our best to find ways, within the framework of our representative form of government, to develop a nuclear waste disposal system that is scientifically, technically, and managerially sound; environmentally and publicly safe; and politically and socially acceptable.

BASIC INFORMATION

This chapter contains basic information about radiation and how it can affect the human body. It also explains the classification of nuclear waste into types, such as high-level or low-level, depending on such criteria as how it was produced and what kinds of radioactivity it contains. Various government agencies bear responsibility for aspects of nuclear waste management; these responsibilities are also described in this chapter.

RADIATION

Because it emits radiation, nuclear waste can harm humans and the environment. Nuclear waste also may contain potentially harmful chemical substances. Whether the potential for harm, or *hazard,* results in actual harm depends on whether people, plants, and animals are actually exposed to radiation and on how much they are exposed to. The *risk* from nuclear waste is measured as a combination of the chance that exposure will occur and the consequences or harm that could result.

Sources of Radiation

Natural radioactive atoms in the earth—primarily uranium, thorium, radium, radon, and potassium—and cosmic rays filtered through the atmosphere from outer space, immerse us in fluctuating amounts of radiation at all times. In addition to this natural background radiation, people are exposed to radiation from manufactured sources. These include medical applications, such as X-rays; consumer goods, such as color television sets and smoke detectors; the operation of the nuclear power industry; the manufacture of nuclear weapons; and fallout from nuclear weapons testing in the past. Of the total amount of radiation that the average person living in the United States is exposed to every year, 82

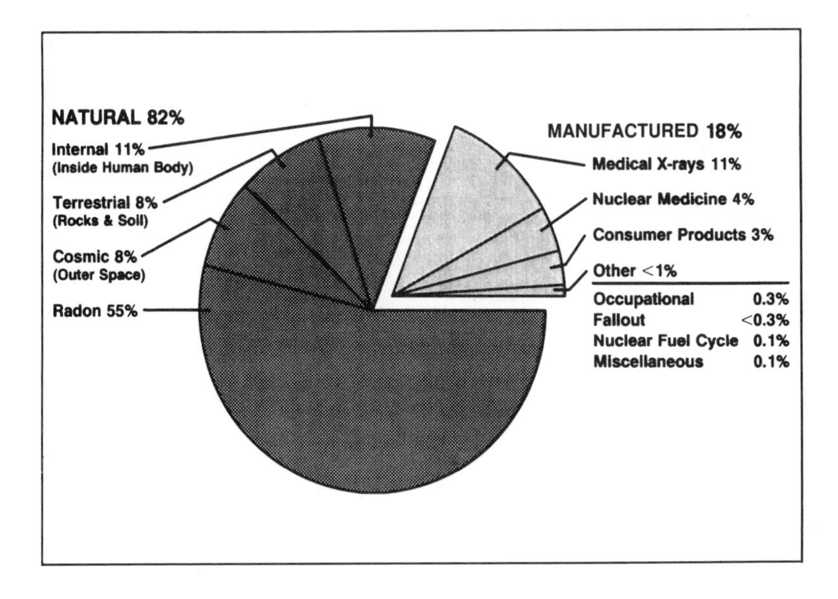

FIGURE 1. Sources of radiation exposure to U.S. population.
Source: National Council on Radiation Protection and Measurements, September 1, 1987.

percent comes from natural sources (55 percent of this is from indoor radon, the importance of which has only recently been recognized), and 18 percent comes from non-natural sources. Medical diagnosis and therapy account for more than 90 percent of the dose from non-natural sources.

Some activities, occupations, and geographic areas expose a person to greater-than-average radiation. For example, a person living at an altitude of 5,000 feet in Denver, Colorado, receives nearly twice as much cosmic radiation from outer space as a person living at sea level in Washington, D.C. Residents in some parts of the country may be exposed to high concentrations of radon from soil.

Most people have received only small amounts of radiation from nuclear weapons production and testing. However, through accidental and planned releases, some employees and neighbors of these facilities have been exposed in the past to radiation doses far higher than would be allowed now.

FIGURE 2: Sample of occupational radiation exposure.

Occupational Category	*Number of Workers (Thousands)*		*Mean annual dose Equivalent[1] (mrem)*		*Collective dose Equivalent*
	All	*Exposed[2]*	*All*	*Exposed[2]*	*(10 person-rem)*
Dentistry	307	61	4	20	1.2
Hospital	140	93	110	160	15.1
Nuclear power reactors	194	100	230	440	44.1
Nuclear fuel fab. & repro.	9	5	80	130	0.7
Air Transport	114	114	350	350	40
Education	17	7	40	80	0.6
Transportation	165	49	8	30	1.4
All Workers	1739	762	85	190	145

[1] These values are rounded to the nearest 10 millirem (mrem).

[2] Workers who received a measurable dose in any monitoring period.

Source: USEPA, *Summary of National Occupational Exposures For 1985*.

RADIOACTIVITY AND FISSION: A SHORT COURSE

Radioactivity. Some atoms, known as radionuclides, are unstable (radioactive) and undergo a spontaneous decay process, emitting one or more types of radiation until they reach a stable form. Called *decay*, this stabilizing process takes from a fraction of a second to billions of years, depending on the type of radionuclide. The rate of radioactive decay is measured in half-lives, the time it takes for half of the radioactive atoms in a sample of a radionuclide to decay to another form. After one half-life, half the atoms in the sample are unchanged; after two half-lives, one-fourth of the original atoms remain unchanged. Thus, after several half-lives only a small fraction of the original radionuclides in the sample remains unchanged, yet the sample may still be quite radioactive. Some atoms may not have decayed and some atoms may have decayed to other radionuclides.

Fission. Some radionuclides are *fissile*; that is, they can split (*fission*) when neutrons are added to their nuclei or, in some circumstances, spontaneously. Uranium-235 is the only fissile nuclide that exists in nature to any extent. Others are produced artificially when *fertile* atoms such as uranium-238 absorb neutrons and subsequently decay to fissile nuclides, such as plutonium-239.

During fission, the nucleus of an atom splits into two smaller nuclei (fission products). Fission products are very radioactive, but they generally lose radioactivity relatively quickly. This splitting also releases neutrons, gamma radiation, and heat. The released neutrons strike other nearby atoms, causing them to split, and, if enough fissionable material is present, an ongoing chain reaction

(continued)

can begin. Such a chain reaction generates heat primarily from the fission process itself and secondarily from the subsequent decay of radioactive fission products. Atomic explosions of nuclear weapons are the result of careful design that quickly brings together a large amount of fissile material so the chain reaction rapidly consumes the fissile material. In a nuclear power reactor, the nuclear chain reaction is controlled so that it cannot produce that kind of atomic explosion. However, destructive but far less intense energy releases are possible, as at Chernobyl, if control of the chain reaction is poorly designed or maintained. In this country, the major control in a nuclear power reactor is in the amount of uranium-235 contained in a reactor's nuclear fuel. Approximately three to four percent of the atoms in commercial nuclear fuel are fissile atoms of uranium-235, compared to 94 percent in a nuclear explosive. In addition, a nuclear reactor contains control rods that absorb neutrons so the chain reaction continues but remains controlled.

Revised and condensed from: *Managing the Nation's Commercial High-Level Radioactive Waste*, Office of Technology Assessment.

Types of Radiation

The radionuclides found in nature, commercial products, and nuclear waste emit three forms of *ionizing* radiation, that is, radiation energetic enough to produce ions by dislodging electrons from an atom. Although all three forms are potentially harmful, they differ in their penetrating power or energy and in the manner in which they affect human tissue.

Alpha radiation is the most densely ionizing but the least penetrating type of radiation. It can be stopped by a sheet of paper. Although alpha particles are unable to penetrate human skin, they may be very harmful if inhaled, ingested,

Radiation Emitted			Radioactive Elements	Half-life		
Alpha	Beta	Gamma		Minutes	Days	Years
☢		☢	Uranium-238			4.5 BILL
☢		☢	↓			
	☢	☢	Thorium-234		24.1	
			↓			
	☢	☢	Protactinium-234	1.2		
			↓			
☢		☢	Uranium-234			247,000
			↓			
☢		☢	Thorium-230			80,000
			↓			
☢		☢	Radium-226			1,622
			↓			
☢			Radon-222		3.8	
			↓			
☢	☢		Polonium-218	3.0		
			↓			
	☢	☢	Lead-214	26.8		
			↓			
☢	☢	☢	Bismuth-214	19.7		
			↓			
☢			Polonium-214	(0.00016 second)		
			↓			
	☢	☢	Lead-210			22
			↓			
☢	☢		Bismuth-210		5.0	
			↓			
☢		☢	Polonium-210		138.3	
			↓			
	NONE		Lead-206		STABLE	

FIGURE 3. The Uranium-238 Decay Chain. ***Source:*** *Radioactive Waste: Issues and Answers*, American Institute of Professional Geologists.

or otherwise admitted into the body, such as through a cut in the skin. Once inside the body, the radionuclide decays, causing highly concentrated local damage. For example, if an alpha emitter were inhaled, the lung tissue could absorb most of the radiation and might become damaged. Long-lived transuranics (elements, such as plutonium, with atomic numbers higher than 92, the atomic number for uranium) emit alpha radiation, as do the decay products of naturally occurring uranium, thorium, and radium.

Beta radiation is a more penetrating type of ionizing radiation. Some beta particles can penetrate skin but, like alpha particles, beta-emitting nuclides may cause the most serious effects if they are inhaled or ingested. Some fission products in spent nuclear fuel assemblies and reprocessed waste—for example, tritium and strontium-90—are beta emitters. The chemical similarities of some of these radionuclides to naturally occurring elements in the body lead them to seek certain organs in the body. For example, the chemical resemblance of strontium to calcium results in its concentration in the bones.

Gamma radiation (high-energy electromagnetic energy waves) has the greatest penetrating power and usually accompanies beta emission. Gamma rays are similar to X-rays—both are electromagnetic radiation—but they have different penetrating power. Gamma radiation from sources outside the body, as well as from sources ingested or inhaled, can penetrate and damage critical organs in the body. Most fission products emit both gamma and beta radiation.

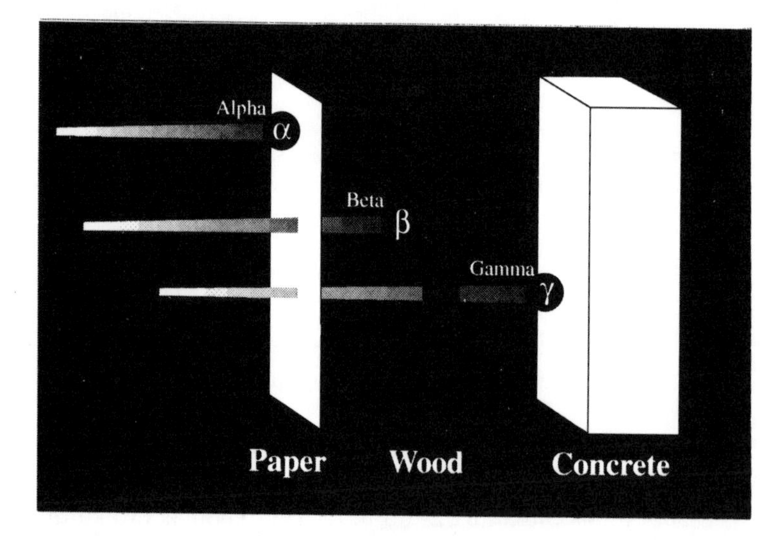

FIGURE 4. Penetrating power of alpha, beta, and gamma rays. ***Source***: SAIC

In high-level nuclear waste, beta and gamma radiation emitters, such as cesium and strontium, present the greatest hazard for the first 500 to 1,000 years; after that, the alpha-emitting nuclides present the greatest hazard. Some alpha emitters have very long half-lives, remaining radioactive for several thousand or even several billion years. Transuranic waste and uranium mill tailings are major sources of alpha radiation. Low-level waste emits alpha, beta, and gamma radiation.

Measures of Radiation

The damage to living material from radiation depends on the amount of energy transmitted to cells and on the number of cells struck. These in turn depend on the type of radiation and on the dose (or total amount of radiation energy absorbed by the struck tissue). Therefore, to assess the possible biological effects of radiation on humans, one measures the amount of energy that has been or might be deposited (in a gram of material, for instance) and the way in which it is or might be deposited.

The *amount*—the radiation dose—may be measured in *rads*. The *biological effect* varies with the type of radiation. For example, a gamma ray may strike several molecules along its path, while an alpha particle creates intense, localized damage. Both of these variables, dose and effect, are roughly taken into account in the measurement called the rem (roentgen equivalent man). The rem is obtained by multiplying the exposure to a certain amount of radiation (measured in rads) by a quality factor that takes into account the differences in the biological damage different types of radiation can do. For X-rays, the quality factor is 1. It is 1 for moderately energetic beta particles, 1.7 or 2 for low-energy betas, and 20 for alpha particles.

Biological Effects

Because the radioactive particles or rays emitted by nuclear waste are energetic, they can cause damage to living organisms. As they travel through human tissue, for example, they strip electrons from the molecules and atoms they strike or pass near, leaving the molecule or atom "ionized" (electrically charged). These ionized particles and the ejected electrons can cause death or damage to cells and cell components and, in sufficient quantity, they can cause the organism to die. The nature and severity of the damage depend on which cell or organ is struck, on the amount of radiation (the exposure) that strikes the body or specific organs or tissues, and on the sensitivity of the struck cell. This same capacity of radiation to alter the structure of cells is used to treat a wide variety of molecular diseases, including cancer.

By studying the records of survivors and victims of the 1945 atomic bombings of Hiroshima and Nagasaki, Japan, scientists have learned a great deal about the biological damage caused by exposure to large amounts of radiation. They are able to predict with some certainty the consequences of exposure to radiation at very high levels. At lower radiation levels, however, the consequences are much more difficult to detect and to predict.

For low radiation doses it is cell damage, not cell death, that is harmful. A dead cell may be replaced, but a damaged cell may replicate itself and multiply. The type of damage resulting from radiation depends on the nature of the struck cell. Damage to an "ordinary" cell, such as bone or organ tissue, is confined to the struck organism. This is called *somatic* damage. The most prevalent type of somatic damage is cancer. Damage to a reproductive cell can cause genetic damage through a *mutation*, transmitting the damage to future generations.

The relationship between low radiation doses—particularly doses below one rem—and the incidence of resulting cancers or mutations is difficult to trace. For both types of damage, the latency period—the time between the exposure and the effect—is long. The latency period can be 25 years or more for cancer and a generation or more for genetic damage. And in both cases, other possible causes—chemical carcinogens, for example—can confuse the issue and make it difficult or impossible to trace the origin of the damage.

Several mathematical methods are used to predict the effect of low-level radiation, but there is disagreement about which method is the most accurate or useful. The controversy arises partly because risk estimates for low levels of radiation are based on sparse data and involve a large degree of uncertainty.

The U.S. Environmental Protection Agency (EPA) establishes standards limiting the allowable radiation dose to the general population, including radiation from nuclear power plants and from other parts of the nuclear fuel cycle. Currently, the EPA limit for the general population is set at a 25-mrem whole-body dose each year from nuclear facilities such as fuel fabrication plants or power plants.

The federal government began setting radiation standards in 1957, and these limits for safely allowable exposure have been made progressively more stringent. The Committee on the Biological Effects of Ionizing Radiation of the National Academy of Sciences, formed at the request of the federal government, in its 1990 report (BEIR V) concludes that cancer risks for a given exposure are about three times larger for solid cancers and about four times larger for leukemia than the risk estimated by the BEIR III committee in 1980. Some people are now calling for EPA to revise its radiation protection standards in light of this finding.

Some nuclear waste also contains hazardous chemical substances. EPA and the Nuclear Regulatory Commission (NRC) are working together to develop regulations for commercial facilities that generate this so-called mixed waste. This is a difficult task. The uncertainty about the biological effects of chemicals is even greater than the uncertainty about the effects of radiation.

To further complicate the matter, chemical constituents found in mixed waste vary, and the complexity of reactions among chemical compounds may change in the presence of radiation. In some cases, the hazardous chemical components of mixed waste may pose the primary health hazard; in other cases, the radioactive components may be the chief hazard; in some, the combination of components may increase or decrease the total hazard of the waste. For example, chemicals may increase or decrease the solubility of a radionuclide and thus influence its ability to move through the environment.

Pathways for Exposure

Toxic material presents a risk to humans or other living things when a person or organism is exposed to the hazard in some way. Radioactive material can reach people and the environment through direct contact or by the movement of radionuclides through the air, soil, surface water, or groundwater. The primary pathways for exposure at a given facility depend on the use and handling of radioactive or hazardous material and on the geology and climate.

Contamination of groundwater by radionuclides is the most probable pathway by which improper nuclear waste disposal can expose humans to radiation. Thus, nuclear waste management and regulation place a great deal of emphasis on preventing groundwater contamination and on detecting and cleaning up groundwater already contaminated by radionuclides.

TYPES AND AMOUNTS OF WASTE

Types

The radioactive nuclides in radioactive wastes differ in the *intensity* of their radiation—that is, by the number and energy of "rays" or particles emitted per second per unit of volume. They also differ in *physical form* (liquid, gas, or solid), in *chemical form*, and in the *type* of radiation they emit (see Types of Radiation, above). Since radiation is a form of energy and since much of the energy released is trapped and remains in the waste material itself, radioactive waste generates heat. The amount of heat generated by a particular kind of waste influences the manner of its disposal.

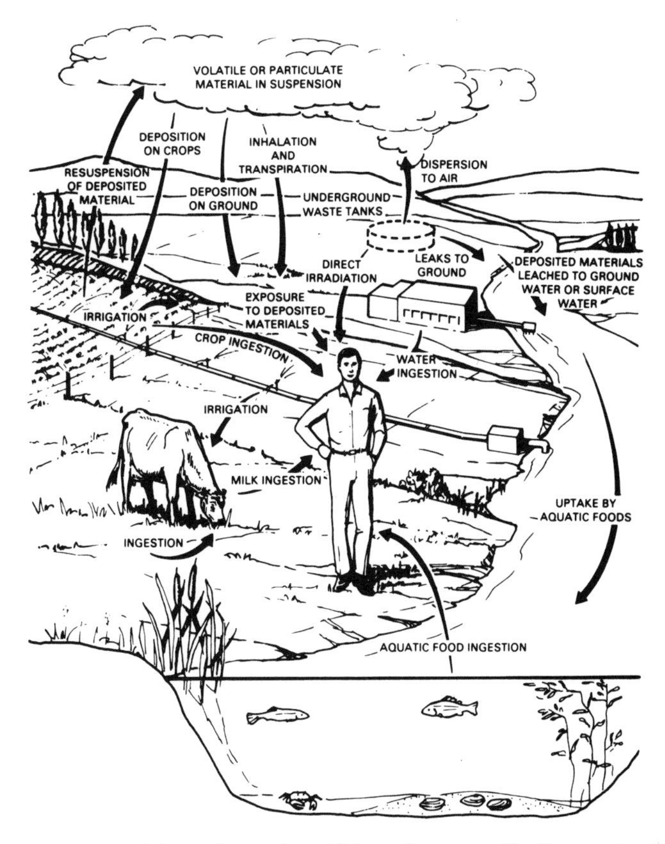

FIGURE 5. Major pathways by which environmentally dispersed radionuclides can affect living organisms. ***Source***: Pacific Northwest Laboratory.

The federal government defines radioactive waste forms as follows:

High-level wastes (HLW) are (1) the highly radioactive waste generated by the reprocessing (chemical separation of the uranium and plutonium from other elements) of used nuclear fuel and (2) the used nuclear fuel itself, if it is not to be reprocessed.

In this country, almost all **high-level waste from repro-**

cessing is the result of reprocessing the used fuel from weapons production reactors in order to obtain material for use in fabricating nuclear weapons. In the United States, as of 1993, this type of high-level waste is mainly stored in liquid form, although some has been physically altered to become a mixture of liquid and sludge or calcine, a dry granular material. All liquid high-level waste will be solidified (turned into glass or other solid form) before it is shipped for disposal. High-level nuclear waste generates much heat and requires heavy shielding to protect humans and the environment from its penetrating radiation. A material such as concrete, water, or lead must be placed between the high-level waste and any person for protection against the danger of external radiation. It must be handled remotely by machines controlled by humans from a distance or behind a barrier.

The **spent fuel** in this country consists mainly of fuel removed from commercial reactors. After three or four years of use, the fuel is no longer efficient in generating electricity. There is also a small amount of special spent fuel from test or research reactors. Spent fuel assemblies are considered a form of high-level nuclear waste because the United States has no present plans to reprocess commercial spent fuel and because the assemblies contain unused uranium, fission products, and transuranic elements, including plutonium. Spent fuel is highly radioactive and generates much heat. It requires heavy shielding and remote handling. After spent fuel assemblies are removed from a reactor, they are submerged in large pools of water to reduce heat and to protect workers from radioactivity. Today, most commercial spent fuel is stored at nuclear power plants in on-site pools. As pools are filled, some partially cooled spent fuel may be moved to massive air-cooled metal or concrete casks for storage. Government-owned spent fuel that will not be reprocessed is stored at DOE facilities in Idaho, Washington State, and South Carolina.

Transuranic waste (TRU) comes primarily from the reprocessing of spent fuel and use of plutonium in the fabrication of nuclear weapons. Transuranic waste is defined by DOE as "waste contaminated with alpha-emitting radionuclides of atomic number greater than 92 [that is, uranium; hence the term transuranic] and half-lives greater than 20 years in concentrations greater than 100 nanocuries per gram." Most transuranic waste emits less intense radiation and generates less heat than fission products, but it typically remains toxic for centuries and requires the same long-term isolation as high-level waste. Alpha-emitting waste generally requires little or no special shielding, but some TRU waste does require protective shielding and remote handling.

Low-level waste (LLW) is a catchall category defined by what it is not rather than by what it is. LLW includes all radioactive waste other than uranium mill tailings, transuranic waste, and high-level waste, including spent nuclear fuel. While most low-level waste is relatively short-lived and has low levels of radioactivity, some presents a greater radiation hazard. Low-level wastes are generated by a wide range of institutions and facilities using radioactive materials, including nuclear power plants, government and defense laboratories and reactors, hospitals, laboratories, and industrial plants. The waste takes a variety of forms, such as medical treatment and research materials, contaminated wiping rags and paper towels, used filters and filter sludge, protective clothing, hand tools, equipment, parts of decommissioned nuclear power plants, and so forth. The radiation from low-level waste sometimes is high enough to require shielding for handling and shipment.

The Nuclear Regulatory Commission classifies low-level waste into four groups according to the degree of hazard it poses and, consequently, the type of management and form of disposal it requires. Low-level waste that can be disposed of by shallow land burial is classified as A, B, or C, from least to most hazardous. States are responsible for the disposal of class A, B, and C waste. The NRC specifies the type of packaging and the form of burial for each of these three classes. Low-level waste that is too hazardous to be disposed of in near-surface facilities is termed "greater than class C waste." Federal law requires that this class of waste be disposed of in a geologic repository licensed for high-level waste or in a disposal facility licensed by the NRC—under more stringent requirements than those for class C waste. The federal government is responsible for the disposal of greater than class C waste.

Uranium mill tailings are the earthen residues, usually in the form of fine sand, that remain after the mining and extraction of uranium from ores. Tailings are produced in very large volumes and contain low concentrations of naturally occurring radioactive materials, including thorium-232 and radium-226. Radium-226 decays to emit the radioactive gas radon-222. Radon is the main radiological health hazard from mill tailings; radon emissions from tailings piles are regulated by EPA.

Mixed waste is waste that contains both hazardous chemical components, subject to the requirements of the Resource Conservation and Recovery Act (RCRA), and radioactive components, subject to the requirements of the Atomic Energy Act. In those states in which EPA implements the entire RCRA Subtitle C program, the hazardous components in a mixed waste management program are regulated by EPA; in those states that have obtained specific authorization from EPA to implement a mixed waste management program, the state regulates the hazardous components. As of January 1992, 29 states had mixed waste authorization. State-implemented RCRA programs may establish more stringent requirements than the federal EPA requirements.

Under the Atomic Energy Act, NRC regulates the radioactive components of mixed waste produced by commercial sources; DOE claims authority to control the radioactive portion of mixed low-level waste produced by the department's weapons production facilities. DOE is the source of about 90 percent of all mixed waste in this country.

EPA, NRC, and DOE are working together to streamline regulations for mixed waste and resolve conflicts among federal regulations. The Federal Facilities Compliance Act, passed by Congress in 1992, requires EPA to develop treatment requirements for mixed waste.

Naturally occurring and accelerator-produced radioactive material (NARM) is yet another category of waste. Accelerator waste, a minor component of NARM, is mainly composed of short-lived radionuclides and is often recycled in an accelerator or used for medical purposes. DOE usually treats accelerator waste as low-level waste. States regulate accelerator waste from private accelerators; most generators store waste on site until it decays. The vast majority of the NARM waste category is composed of naturally occurring radioactive material (NORM). These materials include radium-226, radium-228, radon-222, and other radioactive elements that exist throughout the earth's crust.

There are two types of NORM waste: discrete and diffuse. Discrete NORM has a relatively high radioactivity concentrated in a small volume—examples are industrial gauges, radium needles used in medical equipment, and resins used to remove radium from contaminated drinking water. Diffuse NORM has a low concentration of radioactivity, but a high volume. Diffuse waste includes residue from industrial processes, such as tailings from metal and uranium mining, coal ash from electrical generating utilities, phosphate waste from fertilizer production, contaminated water and drilling equipment from oil and gas production, and sludge from drinking water treatment. Although most diffuse NORM is stored where it is generated, small amounts have been put to use as construction backfill and road building material. Some NORM-contaminated pipes from oil and gas drilling have been reused by schools and other organizations in playground equipment, welding material, and fencing.

The presence of NORM has been known for decades, but only recently have scientists and public officials become concerned about its health effects, realizing that some human activities can concentrate NORM so that it poses a potential health threat. As of 1992, the only NORM subject to federal regulation is uranium mine waste and phosphate waste from fertilizer production. EPA is assessing the extent of the NORM waste problem and is considering regulating it under such acts as the Toxic Substances Control Act and the Resource Conservation and Recovery Act. States are taking the lead in regulating NORM waste. Louisiana has regulated disposal of NORM waste from oil and gas drilling, and other oil and gas producing states may soon follow. The Conference of Radiation Control Program Directors (CRCPD) has developed draft model NORM regulations that states may adopt. One commercial disposal facility, located in Tooele County, Utah, is accepting NORM waste at this time.

Amounts

Existing and projected amounts of waste can be measured either by their volume or by the intensity of their radioactivity. These amounts are shown in Figures 6 and 7.

RESPONSIBILITIES

Federal and state governments bear the vast majority of regulatory and management responsibilities for nuclear waste, but others, such as local governments and industries, also have roles. The following sections outline these responsibilities.

Federal Government

Department of Energy (DOE). Formed in 1977, the Department of Energy absorbed from its predecessor agencies, the Atomic Energy Commission (AEC, 1946-1974) and the Energy Research and Development Administration (ERDA, 1974–1977), the responsibility for various aspects of researching and planning the country's energy supply and delivery, including nuclear power. DOE also is responsible for developing and manufacturing nuclear weapons. DOE carries out federal policies on high-level radioactive waste management and provides technical assistance and coordination to states and other entities on low-level waste management. Different offices within DOE carry out its various functions.

FIGURE 6. Current and Projected Quantities of Radioactive Waste and Spent Fuel

Types of Waste	*Volume of Waste measurements in thousand of cubic meters*				
	1991	***2000***	***2010***	***2020***	***2030***
• High-Level Waste (HLW)					
Commercial*1	1.729	.240	.240	.240	.240
Defense	395	333.6	335.3	341.8	346.4
• Transuranic (TRU)					
Commercial	0	0	0	0	0
Defense*2	254	275	297	191	191
ER TRU	n/a	570	1,100	1,700	1,700
• Spent Fuel					
Commercial*3	9.546	17.091	24.589	31.119	35.353
Defense	0	0	0	0	0
• Low-Level Waste					
Commercial	1,423	1,722	2,055	2,321	2,508
D&D LLW	—	—	7.83	613.06	1,293.30
Defense	2,816	3,787	4,769	5,469	6,231
ER LLW	n/a	920	18,000	29,000	29,000
• Mixed LLW					
Commercial	n/a	n/a	n/a	n/a	n/a
Defense	101.4	n/a	n/a	n/a	n/a
• Uranium Mill Tailings					
Commercial	118,400	119,400	n/a	n/a	n/a
• ER By-product Materials*4	11,390	33,000	36,000	38,000	38,000

n/a Information not available.

ER Waste from DOE environmental restoration projects.

D&D Wastes from decontamination and decommissioning of commercial nuclear reactors.

*1 By the year 2000 HLW is projected to be vitrified into glass, reducing the volume.

*2 Reduction in volume is projected as TRU waste is assayed and some of it is reclassified to other waste categories.

*3 Projected volumes assume no new orders case. Volumes for future years assume the same ratio between the volume and MTIHM of spent fuel as in 1991.

*4 Includes mill tailings, windblown contaminated soil, stabilizing material, low-level waste and source material.

Source: USDOE, *Integrated Data Base 1992: U.S. Spent Fuel and Radioactive Waste Inventories, Projections, and Characteristics*, (DOE/RW-0006, Rev. 8), pp. 15, Table 0.4.

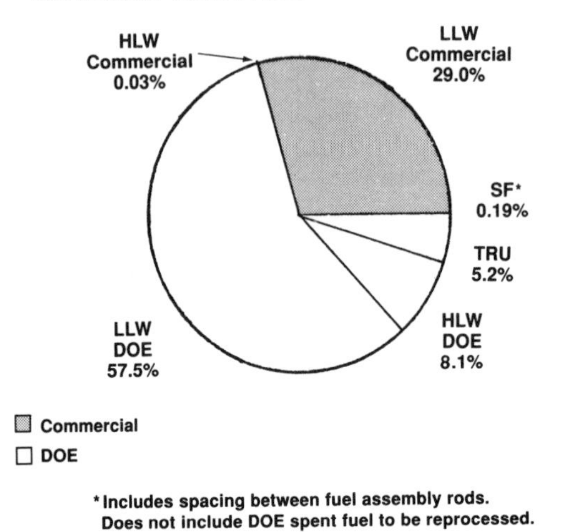

Total volumes of commercial and DOE wastes and spent fuel through 1991.

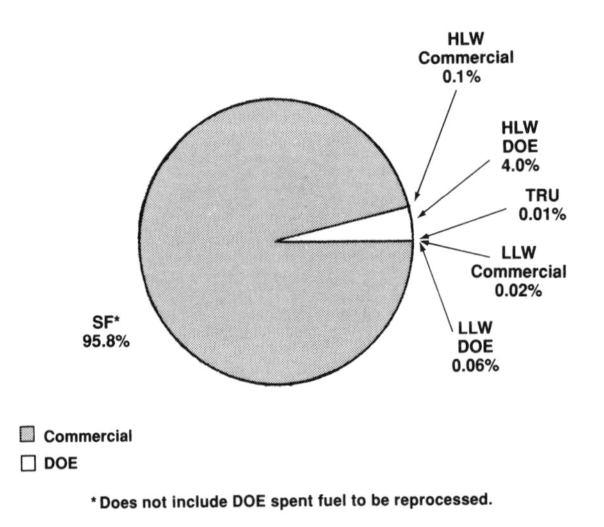

Total radioactivities of commercial and DOE wastes and spent fuel through 1991.

FIGURE 7. A comparison of volume and activity of radioactive waste by source and by type. ***Source:*** USDOE, *Integrated Data Base 1992: Spent Fuel and Radioactive Waste Inventories, Projections, and Characteristics*, (DOE/RW-0006, Rev. 8), page 9, table 0.4.

DOE's Office of Civilian Radioactive Waste Management (OCRWM) is responsible for developing, constructing, and operating one or more geologic repositories for permanent disposal of spent fuel from commercial nuclear power plants and high-level waste from government nuclear weapons facilities. It also is responsible for operating the nuclear waste management system for civilian high-level waste, including transportation and interim storage of spent fuel.

DOE's Office of Environmental Restoration and Waste Management (EM) is responsible for cleaning up or stabilizing DOE sites that have been contaminated by nuclear weapons production and for bringing the operation of federal weapons facilities into compliance with federal environmental laws. It also is responsible for decommissioning (that is, shutting down, cleaning, and monitoring) surplus DOE weapons facilities, managing uranium mill tailings at inactive mills, and treating, storing, and disposing of all categories of wastes generated by DOE weapons production, research, and cleanup activities.

Environmental Protection Agency (EPA). Established in 1970, the Environmental Protection Agency's mandate includes setting generally applicable environmental standards and providing guidance concerning all radiation that affects public health and the environment. EPA develops environmental protection criteria for the handling and disposal of all radioactive wastes. These criteria include limits on radiation releases to the environment and limits on radiation exposure to humans.

EPA also has authority to regulate DOE weapons production sites under the 1976 Resource Conservation and Recovery Act (RCRA) and the 1980 Comprehensive Environmental Response, Compensation, and Liability Act (CERCLA), popularly known as Superfund. EPA, or states under authority delegated by EPA, can regulate the hazardous chemical portion of mixed waste at weapons production sites and enforce requirements of the Clean Water Act, the Clean Air Act and other environmental laws. (See Chapter 6 for an explanation of these laws.)

Nuclear Regulatory Commission (NRC). Established in 1974 when the Atomic Energy Commission was dismantled, the Nuclear Regulatory Commission is an independent federal agency that develops and enforces regulations to assure that public and worker health and safety are protected from all civilian nuclear activities, including those at active uranium mill tailing sites. The NRC has developed regulations for high-level and low-level waste disposal and is responsible for licensing nuclear waste facilities, including a high-level waste repository. It licenses and regulates commercial power plants, industrial firms, individuals, and organizations that possess and use radioactive materials. The NRC shares with the Department of Transportation the responsibility for developing and enforcing safety standards for the transportation of radioactive waste.

Department of Transportation (DOT). The Department of Transportation regulates the shipment of all privately owned radioactive material, including nuclear waste, by all modes of transport. DOT is responsible for developing regulations covering the labeling, classification, and marking of all radioactive waste packages.

U.S. Geological Survey (USGS). The U.S. Geological Survey, in the Department of the Interior, serves as a technical advisor to DOE and the NRC. The USGS has conducted geologic and hydrologic investigations on existing low-level waste sites. Under a Memorandum of Understanding with DOE, the Survey conducts geologic investigations for DOE's high-level waste repository program to provide data for DOE's use in a potential repository license application to the NRC.

Bureau of Land Management (BLM). The Bureau of Land Management, also in the Department of the Interior, reviews environmental assessments, environmental impact statements, land acquisition procedures, and any other plans to site waste management facilities on federal lands over which it has jurisdiction. BLM has provided DOE access to Yucca Mountain, Nevada, for high-level waste repository site investigation.

Office of the Nuclear Waste Negotiator. This office was established by Congress in 1987. Its term now extends until 1995. The Nuclear Waste Negotiator, appointed by the President and confirmed by the Senate, is charged with seeking a state or Indian tribe willing to accept a high-level waste repository or a temporary spent fuel storage facility at a technically qualified site within its jurisdiction. The negotiator is to negotiate the terms of a possible agreement with any state or Indian tribe that expresses an interest in hosting such a facility and to submit any proposed agreement to Congress for approval.

Other

States. Under the Low-Level Radioactive Waste Policy Act of 1980 (see Chapter 4), each state is responsible for ensuring that disposal capacity is available for low-level nuclear waste generated within its borders. The act encourages groups of states to negotiate agreements or compacts to develop disposal capacity jointly. States may receive authority from the NRC to regulate low-level waste disposal facilities (Agreement State status) and authority from EPA to regulate hazardous waste under the Resource Conservation and Recovery Act. Some states have entered into binding agreements with DOE and EPA for cleanup of defense sites; these agreements incorporate state oversight authority over hazardous contaminants. In addition, many states play a role in regulating the transportation of hazardous material, including nuclear waste.

Under the Nuclear Waste Policy Act of 1982 (see Chapter 3), DOE must consult with a state in which it is investigating a site for a high-level waste storage or disposal facility. A state also has the right to disapprove the siting of a high-level waste repository once a site has been selected by DOE and recommended to Congress by the President. A state's veto can be overturned by a vote of both houses of Congress. Under federal law, DOE provides grants to a state to oversee DOE's work when the department is investigating a potential site for a repository or storage facility in the state. DOE also provides grants to a state to assess and mitigate the social and economic impacts of site investigation.

Tribes. Treaties, case law, executive orders, and agreements between the federal government and Native American tribal governments guarantee specific rights to the tribes and provide the basis for a government-to-government relationship. In addition, under the Nuclear Waste Policy Act of 1982, tribes that are designated by the Secretary of Interior as affected by a potential site have the same rights as affected states.

Local governments. Under federal legislation, DOE provides oversight grants to local governments that it has designated as affected by a potential high-level waste storage or disposal facility. The law also stipulates that DOE must make the following payments to a local government whose jurisdiction contains a repository or monitored retrievable storage facility site: payments equivalent to the taxes that would have been paid if the eligible jurisdiction could tax federal activities at a repository or storage facility; impact assistance payments necessitated by the facility siting, construction, and operation; and payment for an on-site representative, designated by the local government in which the site is located, to oversee DOE activities at the site.

Most states' procedures for siting low-level waste facilities provide that any community being considered for a low-level waste facility will be offered a benefits package or may negotiate with the state or facility operator for a benefits package to offset potential negative impacts of the facility. Under some state laws, communities can negotiate for a measure of local control over facility monitoring, emergency closure, and operations such as the types or quantities of waste to be accepted. Communities also may be entitled to receive grants to perform evaluations of potential impacts or to obtain technical advice.

Nongovernment. Utilities, industries, and medical and research institutions that use radioactive material are responsible for some aspects of nuclear waste management, and some play a significant role in formulating and carrying out public policy. Users of radioactive material are responsible for managing the radioactive waste they generate while it is in their possession, although the methods they use to handle, store, and transport the waste are covered by government regulations. Utilities that generate electricity with nuclear reactors collect money from their ratepayers to fund the commercial high-level waste management program.

Under contract and supervision of DOE, private firms operate the department's nuclear weapons facilities. As of September 1992, the department directly employed only 19,000 people nationwide, but DOE's South Carolina Savannah River site alone employed approximately 22,000 contractor employees.

CIVILIAN HIGH-LEVEL WASTE

Since the 1950s, commercial nuclear power reactors have generated spent fuel, a high-level waste, as a by-product of producing electricity. However, the United States has not yet built or even sited a permanent disposal facility for spent fuel, and most current storage facilities were never intended to house spent fuel for more than a few years. The federal government is now investigating a site at Yucca Mountain, Nevada, for a permanent high-level waste disposal facility, or repository. It is also looking for a site at which spent fuel could be stored temporarily, after it has left the reactors and before it reaches the final disposal facility. Many of the people who are looking for solutions to the waste management problem believe that we have sufficient scientific knowledge and technical capacity to site, construct, and operate safe long-term storage and permanent disposal facilities, but that so far we have lacked the political ability or the will to do so. Others involved believe that we still lack the knowledge and technology necessary to develop a long-term plan for disposing of spent fuel.

This chapter explains how high-level waste is produced

FIGURE 8. Civilian Radioactive Waste Facilities. ***Source***: Adapted from *USDOE, OCRWM Bulletin, Spring 1993*.

and stored, discusses some alternative approaches to disposal, and describes the relevant legislation, programs, and issues.

NUCLEAR REACTOR FUEL CYCLE AS A WASTE SOURCE

More than 20 percent of this nation's electricity is produced by nuclear power reactors. Nuclear waste is produced not just at the power plants themselves, but at each stage in the commercial nuclear fuel cycle, beginning with the mining of uranium ore.

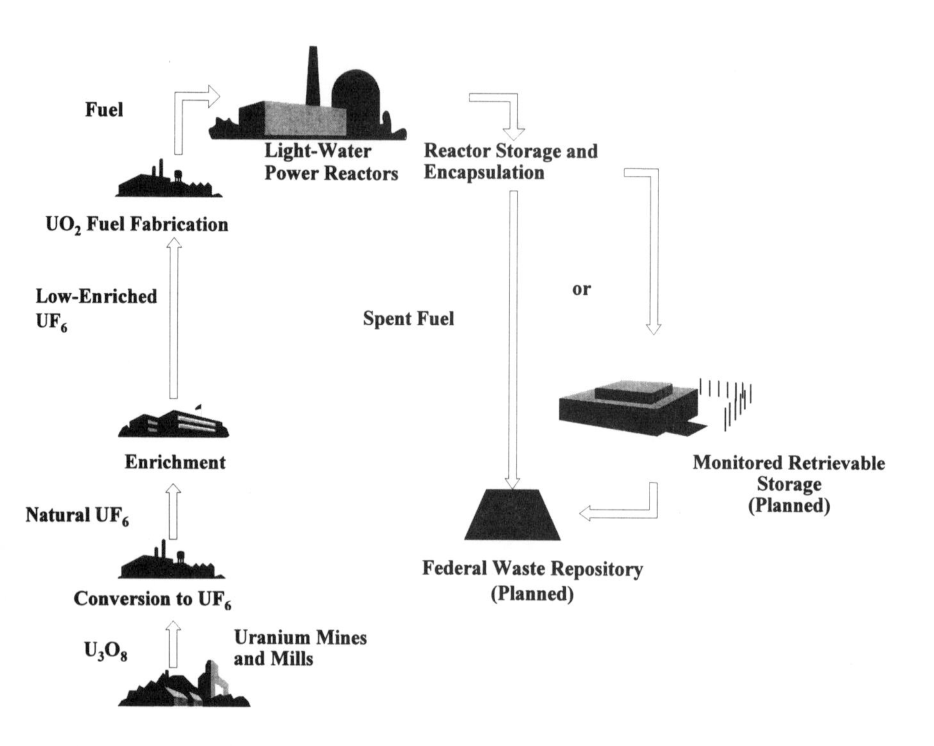

FIGURE 9. The nuclear fuel cycle without reprocessing of spent reactor fuel. ***Source:*** APEX Technology, Inc.

This fuel cycle produces the largest amount of waste, measured by both volume and radioactivity, of all civilian activities. The following section briefly describes the operations that take place at each stage and the wastes that result.

Uranium mining. Routine ventilation of mines results in the release of radon gas and uranium-bearing dust.

Milling. Uranium ore is crushed, ground, and chemically processed to produce the compound uranium oxide (U_3O_8,) known as "yellowcake" because of its bright yellow color. This operation releases small amounts of radon gas and uranium dust. After the refining process, the residue (called tailings) is pumped in slurry form into a settling pond. The water gradually dissipates through seepage and evaporation, leaving behind huge piles of relatively dry, finely ground tailings that contain thorium-230, which decays into radium-226 and then to radon and other long-lived radionuclides. Tailings from mining and milling are generated in huge amounts in this process.

Conversion and enrichment. Yellowcake is converted to uranium hexafluoride (UF_6). Depending on the technique used, the process produces wastes that are either mostly solid or a sludge, with a small part discharged as gas. These wastes contain mainly radium and some uranium and thorium. With the application of heat, UF_6 becomes a gas that permits the concentration (enrichment) of uranium-235, the uranium isotope required for reactor fuel. In this process, small quantities of radioactive gas are vented directly into the atmosphere, and some liquid waste from cleanup operations is diluted and discharged into the environment. Depleted uranium, a by-product of enrichment, is still radioactive enough to be the subject of DOE and NRC regulations.

Fuel fabrication. Enriched UF_6 gas is converted chemically to solid uranium dioxide (UO_2), which is formed into ceramic pellets that are covered with a zirconium alloy, a process known as cladding, to make fuel rods. These are bundled into fuel assemblies containing 50 to 300 rods. Radioactive wastes resulting from these operations include gases and liquid waste containing small quantities of uranium and thorium.

Power plant operation. As the uranium-235 fuel in a nuclear power reactor undergoes fission, producing heat that generates steam for electric power production, the fission fragments or products accumulate within the fuel and gradually reduce the efficiency of the chain reaction. After about three or four years, spent fuel rods are removed from the reactor. Almost all spent fuel rods are now being stored underwater in large pools at reactor sites or, to a lesser extent, in heavily shielded air-cooled casks. Other radioactive wastes generated at nuclear power plants include fission product gases, such as krypton and xenon; carbon-14, mostly as CO_2, from damaged or defective fuel rods; filter media, left over from treating contaminated cooling and cleaning water; and miscellaneous solid waste, such as protective clothing and cleaning paper.

In many other countries, the next step in the fuel cycle is to separate the uranium and plutonium from the fission products in the spent fuel by a chemical process called reprocessing, in order to use the uranium and plutonium again. This process also produces large amounts of radioactive waste. At present, commercial spent fuel is not reprocessed in the United States (see: *Reprocessing and the Nuclear Fuel Cycle*).

SPENT FUEL MANAGEMENT

Most civilian high-level nuclear waste in the United States is in the form of spent fuel and is now stored in pools near the reactors that produced the waste. Some of these pools are nearly full. The following section explains storage practices of the past, present storage and disposal options, and alternative approaches to disposal.

Storage

Storage facilities for spent fuel rods were designed on the assumption that spent fuel would be stored underwater at reactor sites for a short time (about five months to three years) and then shipped away for reprocessing and final disposal. Such a system has not materialized, and most spent fuel remains stored at reactor sites.

By the early 1970s, about 515 tons—six percent of the com-

REPROCESSING AND THE NUCLEAR FUEL CYCLE

The commercial nuclear power system that exists today in the United States is dominated by one kind of reactor, the light-water reactor (LWR), and by a fuel cycle based on "once-through" uranium use. Once-through means that only fresh uranium oxide fuel is used; spent fuel, rather than being reprocessed and used again, is stored until a method of permanent disposal is established.

The once-through cycle uses uranium fuel in a form that cannot be used easily for nuclear weapons. The fission of the uranium fuel in light-water reactors creates plutonium, some of which undergoes further fission and helps generate energy. If the remaining plutonium is never separated from the fuel by reprocessing, it never appears in a form accessible for nuclear weapons. To limit the proliferation of nuclear weapons globally, both President Ford and President Carter imposed indefinite bans on commercial reprocessing in this country, although other nations did not follow the U.S. initiative. In 1981, President Reagan lifted the U.S. moratorium on reprocessing of commercial spent fuel. However, U.S. private industry has no plans to pursue reprocessing because of unfavorable economics, uncertainty about future government policies, and the worldwide abundance of uranium for fabricating reactor fuel. Reprocessing also increases worker exposure and creates significant quantities of liquid high-level waste, transuranic waste, and low-level waste.

mercial spent fuel existing at the time—had been shipped and "temporarily" stored in deep water pools at reprocessing plants in West Valley, New York, and Morris, Illinois. The West Valley facility did reprocess some commercial spent fuel rods before it closed in 1972, but because of design problems, the Morris plant never operated. A third reprocessing plant was constructed at Barnwell, South Carolina, but it was never used because of design and political problems.

Except for 125 spent fuel assemblies owned by DOE, the spent fuel that was stored at the West Valley facility in anticipation of reprocessing has been returned to the nuclear power plants from which it originated. The Morris site is still being used for spent fuel storage. However, the storage capacity of the Morris facility—approximately 720 metric tons—is fully committed under existing contracts with three utilities.

Most spent fuel rods are being stored in pools at reactor sites around the country, and some of these pools are nearly full. According to a 1992 DOE report, 26 reactors will require storage expansion beyond the maximum capacity of their current storage pools by the year 2000. To deal with this problem, most utilities have increased their pools' storage capacity, a step that must be licensed by NRC, by reracking fuel-assembly storage modules to the greatest extent possible. A few power plant operators have moved spent fuel from crowded pools to less-crowded pools at other reactor sites. Some utilities are now using NRC-licensed dry storage technologies. The most fully developed storage technology uses heavily shielded, air-cooled storage casks. Multipurpose casks that can be used to store and transport spent fuel are also being developed. One reactor site in Virginia, one in Maryland, two in South Carolina, and one in Colorado were using storage casks by the end of 1992, with other facilities planning to add dry storage. More than 2,000 storage casks are in use in Europe and Canada.

Disposal

Geologic Disposal

Current federal policy on high-level waste disposal calls for building at least one geologic repository to house the nation's

FIGURE 10. Photo of spent fuel in pool storage at a nuclear power plant. *Source:* USDOE.

FIGURE 11. Photo of dry cask storage system at a commercial nuclear facility. *Source:* USDOE.

high-level waste permanently. As the following section on policies and programs explains in more detail, Congress in 1987 directed DOE to confine its siting investigations for this facility to Yucca Mountain, Nevada. If constructed, the repository would isolate nuclear waste in a stable geologic (rock) formation at least one thousand feet below ground. A combination of natural geologic features and engineered components is expected to provide a series of barriers to prevent the uncontrolled release of radionuclides into the environment. The barriers will include the chemical and physical form of the waste; the covering (cladding) on the fuel rods; the canister that will hold the waste; any packing material around the canister; and the natural characteristics of the rock formation itself.

The concept of geologic disposal of high-level waste and spent fuel has widespread international acceptance in much of the scientific community. A 1992 report from the National Academy of Sciences notes that most countries have concluded that "the best means of long-term disposal of high-level radioactive waste is deep geological emplacement, always including some form of engineered containment or encapsulation and generally with some limited retrieval capability, at least initially."

Geologic disposal has been the focus of federal research for more than 30 years. As early as 1957, a National Academy of Sciences report to the Atomic Energy Commission recommended the burial of high-level and transuranic waste in geologic formations. The Academy urged the investigation of a large number of potential sites and specifically recommended further research on salt beds and salt domes.

In addition to investigating salt extensively (see Chapter 6), DOE has conducted research on geologic formations of basalt, tuff, and crystalline rock (granite) as potential nuclear waste disposal sites. The department conducted experiments in basalt at the Hanford Reservation in Washington and in granite and other kinds of rock formations at the Nevada Test Site, in addition to participating in international research projects.

Shale, alluvium, and argillite formations also have been considered in more preliminary studies. Unlike salt, these types of geologic formations are always fractured to some degree. If these cracks are connected to one another, groundwater can pass through the rock. Shale, alluvium, and argillite (and, to a more limited extent, basalt and granite) formations have the ability to hold onto chemically or to adsorb some waste elements. Thus, if moving groundwater were to leach wastes from a repository and carry it through the rock formation toward aquifers or toward the surface, the ability of the rock to adsorb some radionuclides would retard their movement and help prevent the contamination of water supplies.

Most studies to date suggest that, in a properly sited repository, the odds are very low that groundwater might leach radionuclides from the wastes and carry them to humans and the environment in health-threatening concentrations.

Of course, there are other ways in which radionuclides might be released into the environment, and these must be considered in estimating the risks associated with a potential repository site. Particularly difficult to predict is the likelihood of an accidental release of radionuclides by future human activities, such as exploratory drilling or mining. DOE siting guidelines and procedures, as well as research projects investigating alternative materials and methods, seek to decrease the risk of such occurrences. The guidelines include requirements to consider the likelihood of valuable minerals in the area that might encourage mining and drilling and to post signs warning future generations of the existence of a hazardous site.

Another subject of debate is how long nuclear waste should be retrievable from a repository in case unexpected problems occur or in case future generations wish to recover the buried material. Current NRC regulations require that waste be retrievable for 50 years after a repository begins operation and that the retrieval be no more difficult than the initial excavation.

Some scientists cite the "Oklo" phenomenon as convincing

evidence in favor of disposing of high-level waste in stable geologic formations. Two billion years ago, natural events operating on a very rich uranium deposit in what is now Gabon, in Africa, led to nuclear fission reactions. This "natural" nuclear reactor produced the same types of wastes as today's reactors. Studies of the site, near a village called Oklo, show that most of the fission products and virtually all of the transuranic elements, including plutonium, have moved less than six feet from where they were formed 20 million centuries ago.

Alternatives to Geologic Disposal

While disposal deep within geologic formations has dominated both scientific and policy discussions, other methods for disposing of high-level waste have been considered. The 1980 DOE Generic Environmental Impact Statement evaluated various disposal methods before designating geologic disposal as the preferred alternative.

Subseabed disposal. The only other alternative that has been actively researched is disposal under deep-sea sediments. In a program that began in 1973 in cooperation with several other countries, the United States investigated the feasibility of burying waste packages in geologic formations beneath the deep-ocean floor. Research focused on certain areas in international waters (more than 200 miles from shore) in the western North Pacific and the North Atlantic. In these areas, the ocean is 3,000 to 5,000 meters deep, the sea floor is flat, and the sediments are thick and uniform over a large area. These areas are very stable geologically, isolated from the rest of the planet, and thought to be virtually bereft of life. Sediments in these areas consist of extremely fine-grained clay that would be expected to adsorb most of the radionuclides in the waste. These sediments are considered the primary barrier to the release of radionuclides into the biosphere. The depth of the ocean in these areas would pose a significant barrier to human intrusion.

Research thus far has not revealed any major flaws in the subseabed concept, but important technical questions remain to be answered. These include: whether or not water flows through ocean sediments; if it flows, at what rate it moves; and what effect the heat generated by waste packages may have on the surrounding sediment. The U.S. research project planned to report on the feasibility of the subseabed concept in the 1990s. However, the program was terminated in 1986 by DOE.

The Organization for Economic Cooperation and Development (an organization of economically developed nations, including the United States, Japan, and countries in Western Europe) coordinated the international research program. In 1988 they found that subseabed disposal was technically feasible, but that before subseabed disposal could actually occur, more research was needed to reduce technical uncertainty and an international system was needed to regulate and manage the disposal process.

Currently, the 1976 Convention on Prevention of Marine Pollution by Dumping of Wastes and Other Matter (also known as the London Dumping Convention and signed by most coastal nations) regulates ocean dumping of radioactive and hazardous waste. Although disposal within the subseabed clays differs from dumping on the ocean floor, member nations agreed that none of them should begin subseabed disposal without further research and that, if a nation were to implement a program of subseabed disposal, the London Convention could provide an appropriate international regulatory regime.

Other alternatives. Several other suggested alternatives appear to be impractical. A proposal to bury canisters of waste in the Antarctic ice sheet was abandoned because of uncertainty about the stability of the ice caps over the thousands of years required for radioactive decay of waste. Similarly, although the idea of rocketing nuclear waste into space was ruled technically feasible by the National Aeronautics and Space Administration, it is no longer under investigation because both the cost of such disposal and the risk of a launch accident are considered too high.

Some people argue that we can safely delay taking action on nuclear waste disposal for 50 or more years while continuing to research disposal methods. They recommend storing spent fuel rods in pools of water or in air-cooled above-ground casks or vaults and maintaining continuous human surveillance until the radioactivity and heat given off by the short-lived nuclides have decreased to more manageable levels. Sweden and France, for example, have incorporated long-term (15–30 years) storage as an element of their waste disposal plans, although they also are proceeding to develop geologic disposal.

Proponents of long-term storage for the United States point out that such a plan would buy more time for the development of disposal options. Others, including some critics of the government's lack of progress on nuclear waste management, see this proposal as a delaying tactic. They say this country must develop a permanent solution now before making further commitments to nuclear power. They also argue that long-term temporary storage may become *de facto* permanent storage and that it is wrong to leave this problem for future generations to solve and finance.

POLICIES AND PROGRAMS

Legislation

Before 1982, there was no major legislation in the United States governing the search for a scientifically, technically, and politically acceptable system for managing high-level waste and spent reactor fuel. First the Atomic Energy Commission, then the Energy Research and Development Administration, and now the Department of Energy have been responsible for radioactive waste management. As administrations, agencies, personnel, and political conditions change, so do the answers to basic management policy questions: Should temporary storage be at reactors or away from reactors? Should the facilities be provided by the federal government or by the utilities? Is reprocessing necessary, economically feasible, or politically wise? Is long-term storage desirable, or would it serve only to delay decisions? By what criteria should a permanent disposal site be chosen? What roles should various levels of government, Indian tribes, and the public play in making waste management decisions?

In late 1982, the picture seemed to become clearer. The Nuclear Waste Policy Act, signed into law by President Reagan in January 1983, provided a framework for making decisions about disposal of high-level waste and spent fuel and assigned responsibility for implementing those decisions. The act was based on broad consensus concerning some issues and compromises aimed at balancing interests on others.

Although some of its provisions were altered by amendments in 1987, the Nuclear Waste Policy Act of 1982 set basic policies concerning:

- ▼ *Geologic Repository Development.* The act gave highest priority to permanent disposal in geologic repositories and set a schedule for siting two high-level waste repositories and for constructing and operating one.
- ▼ *Storage.* The act authorized provisions for a limited amount of emergency interim storage and for developing a proposal to site and construct a monitored retrievable storage (MRS) facility on a firm schedule.
- ▼ *Intergovernmental Relations.* The act set requirements for interactions between the federal government and states, local governments, and Indian tribes.
- ▼ *Other Federal Responsibilities.* The act assigned the responsibility for nuclear waste management to specific federal agencies.
- ▼ *Waste Fund.* The act required the establishment of a fund to cover nuclear waste disposal costs paid for by user fees on electricity generated by nuclear power.

Five years later, faced with controversy over some DOE decisions and with concern growing about the slow process and increasing cost of finding a site, Congress passed the Nuclear Waste Policy Amendments Act of 1987, significantly revising the 1982 policies. The 1987 act:

- ▼ Directed DOE to characterize a site at Yucca Mountain, Nevada, to determine whether it is suitable as a repository site, to cease all other repository siting activities, and to postpone consideration of the need for a second repository until the year 2007;
- ▼ Authorized the siting, construction, and operation of a monitored retrievable storage facility subject to certain conditions that link the operation of the MRS very tightly to the construction of a repository;
- ▼ Provided financial incentives for states or Indian tribes on whose land a repository or MRS is sited;
- ▼ Increased external oversight by establishing the Nuclear Waste Technical Review Board, authorizing on-site oversight representatives of host states, Indian tribes, and localities, and providing for increased local government participation;
- ▼ Established the Office of the Nuclear Waste Negotiator to attempt to reach an agreement with a state or Indian tribe willing to host a repository or MRS facility.

Disposal and Storage Siting Efforts

Siting high-level waste facilities has not been easy for a number of reasons, ranging from national policy shifts to states' resistance and local communities' concerns.

Disposal

After studies of many other sites and repeated policy changes summarized below, efforts to site a high-level waste repository have narrowed to a single site at Yucca Mountain, Nevada.

First repository. In 1970, the Atomic Energy Commission tentatively selected a full-scale repository site in the salt deposits near Lyons, Kansas. The site was chosen without a formal search, mainly because it had been used in Project Salt Vault, a test of the effects of heat on salt caverns. Under considerable political and technical fire, the Lyons site was abandoned two years later because of concern that nearby salt mine drilling had compromised the geologic formation's integrity. The government then switched program emphasis from finding suitable underground conditions to developing engineered above-ground structures for storing the waste for an extended period.

But after strong objections that such storage facilities might become *de facto* permanent repositories, in 1974 the federal government again began a search for possible permanent repository sites, beginning with a survey of underground rock formations in 36 states.

In February 1983, following the passage of the 1982 Nuclear Waste Policy Act, DOE formally identified nine potentially acceptable sites located in Louisiana, Mississippi, Nevada, Texas, Utah, and Washington. In draft environmental assessments issued in December 1984, the department recommended further study of sites at Yucca Mountain, Nevada; Deaf Smith County, Texas; and Hanford, Washington.

After President Reagan approved the recommendation of these three sites, DOE began work in 1986 to prepare site characterization plans and establish working relationships with the host states. Although a few local groups welcomed the prospect of site characterization for potential economic benefits, all three state governments opposed the study of sites within their states.

Second repository. The Nuclear Waste Policy Act of 1982 also required DOE to identify a site for a second high-level waste repository. Although not explicitly stated in the act, the intent of the requirement appeared to be to provide some regional equity, with the understanding that the first repository was likely to be in the West while the second would be in another part of the nation.

Indeed, the search for a second site centered on granite formations in 17 eastern, southern, and midwestern states.

DOE issued Draft Area Recommendation Reports in February 1986 and held hearings to discuss the reports throughout the 17 states. Most of the hearings were contentious and packed with citizens who were well organized, well informed, and dead set against further consideration of areas in their states. Governors and members of Congress were alarmed at the uproar, and in May the Secretary of Energy announced that the department was "indefinitely deferring" the "second round" repository program, thus upsetting the act's tenuous regional balance. DOE maintained that, due to a decrease in the projected growth of nuclear power, the second repository would not be needed on the schedule established in the Nuclear Waste Policy Act.

Subsequently, the irate "first round" states allied with the still-uneasy "second round" states to eliminate all funds in the 1988 federal budget for studying the first round sites and for siting the second repository. The siting program ground to a halt as Congress sought a solution to the impasse. The legislated compromise, the Nuclear Waste Policy Amendments Act of 1987, established the potential financial benefits for a host state, terminated all work in Texas and Washington and on second repository siting, and specified that the site at Nevada's Yucca Mountain would be the only one studied.

Characterizing a single site: Yucca Mountain. Not surprisingly, the state of Nevada strongly objected to being singled out by what they viewed as a political rather than a scientific process. The state passed what it considered a legal notice of disapproval of the site under provisions of the Nuclear Waste Policy Act (see Intergovernmental Relations below) and refused to issue permits necessary for DOE to begin site characterization. The U.S. Court of Appeals ruled and the U.S. Supreme Court upheld that the notice was premature; that is, the state could officially disapprove only when site characterization was complete and when the President had recommended to Congress that the site be developed as a repository. The state agreed to process the necessary permits.

In 1987, as required, DOE submitted a site characterization plan to the state and to the NRC and held public hearings. In 1991, the department began conducting surface studies at the site and preparing to construct a facility for underground tests and exploration.

The site at Yucca Mountain is in tuff, a rock formed of compacted volcanic ash. Unlike other proposed sites, Yucca Mountain is in an unsaturated zone; that is, it is above the current water table.

The Nuclear Waste Technical Review Board was established by Congress to provide technical oversight of the repository studies. Its members are appointed by the President from a list of scientists and engineers nominated by the National Academy of Sciences. Issues that the Technical Review Board has stressed for study include seismic vulnerability (the likelihood that adverse consequences will result from earthquake ground motion or fault displacement) and the role that engineered barriers, especially very durable canisters, can play in providing additional protection. The review board maintains that these issues can be resolved only by underground studies at the site.

Issues of particular concern to the state and to local communities in Nevada are groundwater movement, faults in the rock, the potential for earthquakes and volcanic activity at the site over the long time period for which the repository must remain secure, and the possibility of negative impacts on tourism and economic development. Furthermore, Yucca Mountain is within a significant mining district, and there is a possibility of deep gas or oil deposits in the area. Thus, state officials and residents are also concerned that people might unknowingly drill or mine into the repository after it is closed.

If at any time during site characterization the department determines the site to be unsuitable for development as a repository, the Secretary of Energy must terminate activities at the site and, within six months, recommend to Congress what alternative action should be taken to "assure the safe, permanent disposal of spent nuclear fuel and high-level radioactive waste."

Once the characterization process is complete (a task expected in 1992 to take seven to ten more years at a total cost of $6.5 billion), if the site is determined to be suitable under the siting guidelines, DOE will prepare an environmental impact statement and send a recommendation to the President that the site be developed as the first U.S. permanent high-level nuclear waste repository. The recommendation will include comments from the NRC and the state of Nevada. If the President agrees that the Yucca Mountain site is qualified, the recommendation will go to Congress. At that time, the state of Nevada may issue a notice of disapproval. That notice, or veto, will stand unless overridden by a joint resolution of Congress.

When a repository site has been designated, DOE must submit a license application to the NRC for construction authorization. The commission must make a decision on the application within three years, though a one-year extension is possible.

EPA is charged with setting environmental protection standards for a repository. The agency issued standards in 1985 setting limits on radiation releases to the general environment and exposure to humans. However, ruling in 1987 on a lawsuit brought by environmental groups and states, the U.S. Court of Appeals for the First Circuit instructed EPA to reconsider some parts of the standards because of inconsistencies with the Safe Drinking Water Act and inadequacies in the standard governing individual exposure. By mid-1992, EPA had circulated for comment a series of drafts of proposed revised standards but had not issued final standards. However, the Energy Policy Act of 1992, passed in October, mandated a new process for issuing standards specifically for a repository at Yucca Mountain. It directs (1) the National Academy of Sciences to form a committee to provide "findings and recommendations on reasonable standards for the protection of public health and safety" no later than December

31, 1993, and (2) EPA to issue standards for the Yucca Mountain site consistent with these findings and recommendations within one year after receiving them.

In 1981, the NRC issued regulations, based on anticipated EPA standards, for a mined geologic repository These standards have been amended several times and will be revised, if necessary, to be compatible with EPA's new standards when they are final.

Storage

The Nuclear Waste Policy Act of 1982 emphasized the responsibility of the owners and operators of civilian nuclear power reactors to provide interim storage by maximizing the effective use of existing storage facilities and by adding new on-site storage capacity. The 1982 act also required DOE to study the need for and feasibility of a facility for the long-term storage of spent fuel (monitored retrievable storage) and to submit a proposal to Congress for the construction of one or more such facilities.

In April 1985, DOE recommended the construction of a monitored retrievable storage facility as part of an integrated waste management system and proposed consideration of three sites in Tennessee. The report described a facility that would receive spent fuel from commercial power reactors, consolidate and package the spent fuel, and then store the furl temporarily, pending shipment to a repository. It would be centrally located near the majority of reactors, and the impact of transportation to the final disposal facility would be minimized by shipping spent fuel in large rail casks on dedicated "unit trains" used only to transport this cargo.

The department's preferred site was near Oak Ridge, Tennessee. Although the city of Oak Ridge concluded that a monitored retrievable storage facility would be acceptable under certain conditions, the state of Tennessee sued to block the submission of DOE's report to Congress, arguing that the act required DOE to consult with the state before choosing a specific site. Eventually, Tennessee's legal petition was denied, but its point had been made. The Nuclear Waste Policy Amendments Act of 1987 revoked the proposal.

The 1987 Nuclear Waste Policy Amendments Act did authorize DOE to site and construct a monitored retrievable storage facility, with strong restrictions. The department cannot select an MRS site until a permanent repository site has been recommended, and construction cannot begin until the NRC has issued a construction license for a repository. Only a limited amount of spent fuel can be stored at any time—spent fuel equivalent to 10,000 metric tons of heavy metal before a repository is operating or 15,000 metric tons of heavy metal when a repository is operating. The act instructed DOE to evaluate the use of dry cask storage at reactor sites in consultation with the NRC.

In addition, the Office of the Nuclear Waste Negotiator, created by the 1987 act, is to "attempt to find a state or Indian tribe willing to host a monitored retrievable storage facility at a technically qualified site on reasonable terms" and to negotiate an agreement with the governor of that state or governing body of that Indian tribe. To become effective, the agreement must be enacted into law by Congress. Responding to an invitation from the negotiator, some counties and Indian tribes have been willing to explore the possibility of siting a monitored retrievable storage facility. As of December 1992, four counties and 16 Indian tribes had applied for grants to study the feasibility of locating a storage facility in their jurisdictions; three counties and seven tribes were awarded grants. However, one county and four tribes subsequently withdrew from the process.

DOE initially decided not to conduct a siting process of its own but to rely on the voluntary process described above to identify a site for an MRS in time for a facility to be operating by January 1998. That date is important because contracts between DOE and utilities specify that DOE will begin to take responsibility for spent fuel beginning in 1998. However, utilities have become increasingly concerned that, without an MRS or repository available by then, the federal government will be unable to meet its commitment. Indeed, according to the department, a volunteer site for an MRS would need to have been identified before October 1992 for all the steps necessary to construct an MRS to be completed before 1998.

Utilities and state rate-setting commissions have pointed out that they are collecting fees from consumers for the Nuclear Waste Fund in exchange for the federal government's assuming the obligation to begin accepting spent fuel in 1998. They also have pointed out that reactor sites had been chosen and licensed for reactors with lifetimes of 40 years, not as sites for indefinite storage of spent fuel.

Responding to these concerns, DOE announced a major change in policy in December 1992. DOE will look for interim spent fuel storage capacity at nuclear weapons facilities and other federal sites that could be ready to receive nuclear utility spent fuel by January 1998. The department also will begin considering ways to compensate nuclear utilities that may have to pay for additional on-site spent fuel storage after January 1998 and before a federal facility is opened.

Intergovernmental Relations

Relations have been less than harmonious between the federal government and the states that either contain identified sites for waste management facilities or fear they may be next on the list. In fact, more than a dozen states, responding to pressure from citizens, have enacted laws intended either to prohibit flatly or to make it difficult to establish within their borders disposal facilities for radioactive waste. However, such prohibitions on nuclear waste facilities may not pass constitutional muster because of conflicts with the commerce clause of the U.S. Constitution.

Why do so many state and local governments want to restrict or prohibit nuclear waste disposal and even temporary storage within their boundaries? One reason is that adverse experiences with other federal and private projects involving hazardous substances have made states wary of possible future problems from nuclear waste facilities. Citizens and state and local officials want assurances that the facilities will be properly constructed and operated, and that they will pose no significant risks to people or to the environment now or in the future. Some want their states to play no part in disposal or storage under any conditions.

Western states feel they have long been targeted for hazardous facilities. These states have been the sites for many federally sponsored hazardous activities in the past, including uranium mining, milling, and tailings disposal; nerve gas production, testing, and storage; and atomic bomb testing. Often, these remote locations were selected to minimize risk to the population at large. But as one Westerner put it, "The government has used the wide open spaces as a dumping ground for almost four decades and inflicted a lot of wounds on us. Well, we've just had enough." On the other hand, some people living near potential sites welcome a nuclear waste repository or storage facility for the economic benefits they hope it will bring.

Both the Nuclear Waste Policy Act of 1982 and Nuclear Waste Policy Amendments Act of 1987 provide some protection for state and local governments. Yet, by narrowing site investigations to Yucca Mountain largely through a political rather than a technical process, Congress also set the stage for confrontation between the state of Nevada and the federal government.

Nevada has refused to discuss entering into a benefits agreement, which under the Nuclear Waste Policy Amendments Act would require the state to waive its right to disapprove the site. The state of Nevada and certain affected localities are receiving grants from DOE to assist them in independent oversight of the program, reviewing the department's work, assessing potential impacts, providing information to Nevada residents, and monitoring, evaluating, and commenting on site characterization activities.

No legislation can guarantee agreement between states and the federal government. Tension is inevitable since state, local, and federal governments have different responsibilities and often different goals.

RADIOACTIVE WASTE ISSUES IN INDIAN COUNTRY

The participation of Indian tribes in feasibility studies for the monitored retrievable storage facility for spent nuclear fuel surprised many people who questioned tribal involvement in something they considered so alien to the tribal experience. However, the MRS study is not the first aspect of nuclear energy and radioactive waste in which Indian tribes have played a role. They have been intimately involved in or affected by nuclear energy and radioactive waste since the very beginning of the atomic era. The Trinity test site, the site of the first human-made nuclear explosion, is not far from the home of the Mescalero Apaches, one of the tribes now studying the MRS. The uranium for nuclear bombs was enriched at Hanford on lands ceded by the Yakima Indian Nation and adjacent to lands in which the Nez Percé Tribe and Confederated Tribes of the Umatilla Indian Reservation have treaty interests. Some of the uranium used to build our nation's nuclear stockpile, fuel ships and submarines of the U.S. Navy, and power the commercial reactors came from mines on land of the Navajo Nation, the Spokane Tribe, and the Pueblo of Laguna.

By the late 1970s and early 1980s, it became apparent that nuclear power was not going to enjoy the kind of expansion that utility and governmental proponents had anticipated. As interest in the exploration and development of tribal uranium resources waned, and as uranium mining and milling in Indian country and in the United States generally ground to a halt, the dark legacy of such exploration and development manifested itself in abandoned and unreclaimed mines, in mine tailings and tailings ponds throughout Indian country, and in increased numbers of Indian uranium miners sick and dying of cancer.

(continued)

Also during the 1970s and early 1980s, the extensive contamination of the enormous nuclear weapons complex came to light. Awareness of this environmental problem and the end of the Cold War changed DOE's mission from building atomic weapons to cleaning up the weapons complex.

Indian tribes today are affected by these changes in our national defense posture and by changes in the energy mix. Following are current examples of tribal involvement in radioactive waste management.

Uranium mining impacts. Uranium mining on Navajo land was conducted in a less-than-safe manner that increased miners' exposure to radioactivity and led to a higher-than-normal incidence of cancer and lung disease among the miners. Federal legislation to compensate the miners and their families was recently enacted. The Spokane Tribe today is faced with an enormous unreclaimed mine and a large tailings pond that pose a threat to the water systems in the region. In general, exploratory digs on Navajo, Pine Ridge, and other Indian reservations have not been reclaimed, and they present health and safety as well as environmental problems.

Transportation of wastes through Indian lands. Indian tribes have regulatory authority on their lands. Spent nuclear fuel and wastes generated by the cleanup of the nuclear weapons complex are and will continue to be transported through Indian lands. This transport requires tribes to develop and implement transportation, emergency response, and health and safety programs. The Shoshone-Bannock Tribes in Idaho, the Confederated Tribes of the Umatilla Indian Reservation in Oregon, and the Acoma Pueblo in New Mexico are among those tribes that have developed some emergency response capability.

Tribal treaty rights on lands occupied by the nuclear weapons complex. Tribes in the Pacific Northwest and in Nevada have or claim treaty rights—including hunting, gathering, and access for religious purposes—on lands currently occupied by the DOE weapons complex. With DOE financial assistance, the Yakima Indian Nation, the Confederated Tribes of the Umatilla Indian Reservation, and the Nez Percé Tribe are involved in the cleanup of the radioactive contamination and other contamination at Hanford.

Tribal sovereignty issues. The question of tribal sovereignty has arisen in the context of the MRS studies by Indian tribes. Even when the underlying concern is environmental protection, health, safety, or the belief that temporary custody of spent nuclear fuel by Indian tribes is somehow contrary to Indian values, the argument generally focuses on whether or not tribal sovereignty should extend to tribal involvement in radioactive waste management.

Transportation issues also have jurisdictional implications. Radiological emergency response capability is extraordinarily expensive to put into place and to maintain. Regional intergovernmental approaches to emergency response planning and program development are encouraged by federal agencies and by the practical realities of the limited financial resources available for building state and tribal capabilities. Unlike states, many tribes lack basic emergency response capabilities and, therefore, will need to enter into regional or local cooperative services arrangements or obtain the funding and technical assistance needed to develop the ability to respond to emergencies. Routing decisions, reciprocity in vehicle inspections, and development of driver standards all have jurisdictional and intergovernmental implications.

—Mervyn L. Tano
Council of Energy Resource Tribes

Other Federal Responsibilities

The Nuclear Waste Policy Act of 1982 established the Office of Civilian Radioactive Waste Management, within DOE, to implement the act, and also required that DOE assess alternative ways of managing the civilian waste management program. In 1984, a committee appointed by the Secretary of Energy recommended that an independent, federally chartered corporation be set up to manage the civilian nuclear waste program. However, an internal DOE review committee rejected that proposal and the Secretary of Energy concluded that the present structure (the DOE Office of Civilian Radioactive Waste Management) should be retained, at least through the siting and licensing stages.

The act gave DOE two other responsibilities: to develop transportation and interim storage plans and to make a final recommendation about whether defense waste should be disposed of in civilian repositories. In April 1985, President Reagan accepted DOE's recommendation to dispose of defense waste in civilian repositories, making such waste subject to NRC requirements.

Waste Fund

The Nuclear Waste Fund, provided for in the Nuclear Waste Policy Act of 1982, is supported by user fees intended to underwrite fully the costs of the DOE civilian radioactive waste management program mandated in the legislation. In exchange for these payments from utilities, the government was required to enter into contracts with the owners and generators of high-level waste to begin accepting waste from utilities for disposal by January 31, 1998.

Based on the assumption that consumers of electrical power generated by nuclear energy should bear waste disposal costs, the fund assesses two kinds of fees: (1) a one-time charge for commercial high-level waste or spent fuel in existence before April 1983 and (2) an adjustable fee, initially one mill (.1 of a cent) per kilowatt hour, levied on electricity generated by nuclear reactors after April 1983. This fee is subject to annual review and adjustments to ensure that it covers all costs. Despite concern by the General Accounting Office and others that the 1 mill is inadequate, DOE has not recommended an increased fee.

By mid-1992, the fund had received a total of $7 billion in fees and interest. The fund is expected to receive annually an additional $600 million from fees (the exact amount will depend on how much electricity nuclear power plants sell to consumers) and $200 million from interest. This money is kept in the U.S. Treasury and cannot be used for other projects. However, Congress must appropriate funds in the federal budget before they can be used. Approximately $3.3 billion has been spent to date. In 1992, DOE estimated that site characterization alone will cost more than $6.5 billion dollars to complete.

WHAT OTHER COUNTRIES ARE DOING

At the start of 1992, 417 nuclear power reactors were operating in 31 countries and were producing 17 percent of the world's electricity. Nearly all of these countries have waste management programs. Although situations vary from country to country, most waste management programs for high-level waste assume that spent fuel will be reprocessed and that the resulting high-level waste will be vitrified into glass and then disposed of in a deep geologic repository. To date, no permanent disposal of high-level waste has taken place in any country. Most countries face political opposition to siting high-level waste facilities; as a consequence, some nations are only researching disposal alternatives and are contracting with other countries for reprocessing services. Standard processes for disposal of low-level waste range from shallow land burial to underground disposal in played-out mines or in near-surface or deep repositories especially constructed for the purpose.

(Continued)

Cooperative International Research Programs

The Nuclear Energy Agency (NEA) is a 23-member agency of the Organization for Economic Cooperation and Development (OECD). The International Atomic Energy Agency (IAEA) is a 113-member independent organization under the aegis of the United Nations. These two agencies coordinate multinational research programs. The NEA facilitates the exchange of information on nuclear waste issues, conducts and sponsors international research and development projects (focusing particularly on performance and safety assessment), and coordinates *in situ* research, site investigations, and underground demonstration projects by its members. The IAEA programs in which the United States participates concentrate on radioactive waste safety standards, safeguards for spent fuel storage and handling, and transportation regulations.

The United States has also signed bilateral agreements to undertake research on nuclear waste management with each of the following countries: Belgium, Canada, the Commission of the European Communities, France, Germany, Japan, the former Soviet Union, Spain, Sweden, Switzerland, and the United Kingdom.

FIGURE 12. High-level waste burial program in other countries. ***Source:*** adapted from "Nuclear Waste: The Problem that Won't Go Away. *Worldwatch Institute*, December 1991, pp. 24–25.

Country	*Earliest Planned Year*	*Status of Program*
Belgium	2020	Underground laboratory in clay at Mol.
Canada	2025	Independent commission conducting four-year study of government plan to bury irradiated fuel in granite at yet-to-be-identified site.
China	none announced	irradiated fuel to be reprocessed; Gobi desert sites under investigation.
Finland	2020	Field studies being conducted; final site selection due in 2000.
France	2010	Two sites to be selected and studied; final site not to be selected until 2006.
Germany	2008	Gorleben salt dome sole site to be studied.
India	2010	Irradiated fuel to be reprocessed, waste stored for twenty years, then buried in yet-to-be-identified granite site.
Italy	2040	Irradiated fuel to be reprocessed, and waste stored for 50–60 years before burial in clay or granite.
Japan	2020	Limited site studies. Cooperative program with China to build underground research facility.
Netherlands	2040	Interim storage of reprocessing waste for 50–100 years before eventual burial, possibly sub-seabed or in another country.
Russia	none announced	Current Russian program uncertain.
Spain	2020	Burial in unidentified clay, granite, or salt formation.
Sweden	2020	Granite site to be selected in 1997; evaluation studies under way at Aspo site near Oskarshamn nuclear complex.
Switzerland	2020	Burial in granite or sedimentary formation at yet-to-be identified site.
United States	2010	Yucca Mountain, Nevada, site to be studied, and if approved, receive 70,000 tons of waste.
United Kingdom	2030	Fifty-year storage approved in 1982; explore options including sub-seabed burial.

FOUR

CIVILIAN LOW-LEVEL WASTE

Civilian low-level radioactive waste results from the use of radioactive materials in such diverse activities as generating power by nuclear reactors, conducting medical or biotechnological research, performing medical examinations and treatment, producing radioactive chemicals for use in nuclear medicine and research, and controlling the quality of products being manufactured. Its various forms include contaminated filters, liquid filter resins, wiping rags, protective clothing, hand tools, vials, needles, test tubes, animal carcasses, lab equipment, luminous dials, sealed radiation sources, and internal components from reactors.

Until the early 1960s, commercial low-level waste was disposed of in federal disposal facilities. When the federal government closed its facilities to commercial waste, private companies opened and operated disposal facilities, some of which encountered problems and closed. By the late 1970s, all low-level waste in the United States was being shipped for disposal to only three states: Nevada, South Carolina, and Washington. Pressured to make other states share the disposal burden, Congress passed the Low-Level Radioactive Waste Policy Act in 1980, making each state responsible for providing disposal capacity for the commercial low-level waste generated within its borders. Congress amended the act in 1985 to give states more time, but progress still has been slow. As of the end of 1992, no state or regional group of states had actually sited, licensed, and constructed a low-level waste disposal facility, though some have tried very hard to do so.

SOURCES AND VOLUMES

Percentages vary considerably from year to year, but in 1991 about 46 percent of the total volume of low-level waste generated in this country came from nuclear power plants, 6 percent from medical and research facilities, 40 percent from industry, and 8 percent from government and military operations.

FIGURE 13. Source of low-level radioactive wastes received by commercial disposal sites by volume and radioactivity.

Generator Type	*Average Yearly Volume (FT^3)*	*Percent of Total*
Academic	51,280	3.51%
Government	107,182	7.33%
Industry	567,753	38.81%
Medical	27,313	1.87%
Utility	709,215	48.49%
TOTAL	1,462,743	100.00%

Generator Type	*Average Yearly Radioactivity (curies)*	*Percent of Total*
Academic	1,504	0.22%
Government	18,659	2.68%
Industry	87,014	12.52%
Medical	153	0.02%
Utility	587,681	84.56%
TOTAL	695,012	100.00%

Source: USDOE, National Low-Level Waste Management Program, 3/3/93.

At the end of 1992, commercial low-level nuclear waste disposal sites at Beatty, Nevada, Barnwell, South Carolina, and Richland, Washington, were still accepting waste. In 1991, the Nevada facility accepted approximately 12 percent of the nation's commercial low-level waste; the South Carolina site received approximately 58 percent; and the Washington site received approximately 31 percent. By the end of 1992, the Nevada site was closing and the Washington site was prepared to accept waste only from a limited number of states, leaving South Carolina as the destination for most of the country's low-level waste in 1993.

Reduction of Low-Level Waste

Public pressure, uncertain access to disposal facilities, and increasing disposal costs for low-level waste have heightened interest in both source reduction (reducing the amount of waste actually generated) and volume reduction (reducing the quantity of waste after it is generated). Many companies and institutions have accomplished significant source reduction in low-level waste by changing manufacturing processes, handling radioactive material more carefully, and segregating waste more precisely. Many also have achieved volume reductions by increased use of compaction, incineration, filtration, and evaporation. By reducing the volume of low-level waste for disposal, these actions can extend the operating life of disposal sites, reduce the disposal capacity needed at new sites, limit the need for interim storage, and reduce the need to transport low-level waste. However, although source reduction may reduce the amount of radioactivity in the waste, volume reduction concentrates rather than reduces total radioactivity. That is, volume reduction increases the number of curies per unit volume of waste.

The volume and the radioactivity of the waste stream result from decisions made by many diverse companies and institutions. Although the overall trend since the early 1980s reflects a reduction in low-level waste volume, the trend since 1985 has been less clear-cut. Volumes have fluctuated and are not expected to decrease significantly; the overall concentration of radioactivity in the waste has increased significantly. The number of nuclear reactors being refueled has a large impact on the volume of waste generated; the number of nuclear reactors being decommissioned will have even more of an impact.

DECOMMISSIONING NUCLEAR POWER PLANTS

Commercial nuclear power plants that have shut down and plant equipment that is no longer used constitute a potentially large source of nuclear wastes. During the fission process that powers a reactor, neutrons bombard not only the uranium fuel but also other parts of a nuclear reactor, including the steel structures that support the fuel, the steel reactor vessel that holds both the fuel rods and the coolant, and the massive concrete containment structure that shields the reactor vessel. During the life of the reactor some neutrons are absorbed by atoms of cobalt, iron, nickel, and other elements in the steel, water, and concrete.

Because some of the resulting nuclides, called activation products, will remain highly radioactive for several decades or more, a nuclear power plant must be closed down, or "decommissioned," in a way that will prevent public exposure to or dispersion of radioactivity. Decommissioning creates a large and varied amount of waste including large quantities of low-level waste, equipment and building materials contaminated by fission products, high-activity wastes containing activation products, and spent fuel.

By early 1993, 16 licensed commercial power reactors

(continued)

will have shut down in the United States. In 1991, DOE estimated that 65 of the nation's 109 operating nuclear power plants will shut down permanently by 2020. Although 60-70 small experimental and prototype reactors have been decommissioned to date, only three small nuclear power-generating reactors have been fully decommissioned in the United States: Elk River Plant, Minnesota; Shippingport, Pennsylvania; and Santa Susana, California. These three reactors were all owned by DOE or its predecessor agency, the Atomic Energy Commission, and all produced some commercial power. At the Pathfinder power reactor in North Dakota, the spent fuel was removed and contaminated areas were decontaminated or secured in 1967 (see SAFSTOR below). Dismantlement began in July 1990 and is nearly complete. Pathfinder is the first privately owned reactor to undergo decommissioning.

Predicting the pace at which commercial nuclear power reactors will be decommissioned has turned out to be more uncertain than the industry expected. Initially commercial reactors are licensed to operate for 40 years; the NRC is developing regulations to extend licenses of plants and allow them to operate even longer. However, some reactors do not complete their licensed operating lifetimes for various reasons, such as regulatory and operating difficulties, economic considerations, and, in the case of Three Mile Island, an accident.

Methods for Decommissioning Power Plants

Three decommissioning approaches are recognized worldwide: immediate dismantlement, safe storage with later dismantlement, and entombment. The Nuclear Regulatory Commission calls these three options DECON, SAFSTOR, and ENTOMB. Before decommissioning begins, the plant owners' decommissioning plans must be approved by the NRC. The choice between immediate or delayed decommissioning involves such questions as whether the cost of storing a reactor while the radiation levels drop are offset sufficiently by the savings in radiation exposure to workers and lower costs of decommissioning a less radioactive plant.

Immediate dismantlement (DECON). In this approach, the power plant is decontaminated soon after shutdown and all radioactive components, solid and liquid, are removed to a radioactive waste storage or disposal facility. After dismantlement is complete, the nuclear license is terminated, and the property is released for unrestricted use.

Two plants have been DECONed in this country and work has begun on the third. Dismantlement of the Elk River Plant in Minnesota, a 22.5-megawatt demonstration reactor, was completed in 1974. Dismantlement of a slightly larger, 72-megawatt power plant at Shippingport, Pennsylvania, was completed in 1982. Fort St. Vrain, a 330-megawatt plant in Platteville, Colorado, closed in August 1989; it has been defueled and will be dismantled. Dismantling Fort St. Vrain will generate an estimated 140,000 cubic feet of low-level waste for disposal. Some of the spent fuel is stored in an on-site dry storage facility; some is being sent to Idaho National Engineering Laboratory to be used for research.

Safe storage with later dismantlement (SAFSTOR). In this approach, after the spent fuel is removed, contaminated areas are either decontaminated or secured. The plant is then isolated to allow for further radioactive decay, primarily of highly radioactive activation products. The structures and equipment to be dismantled later are securely maintained to protect the public from exposure to residual radioactivity. Once the plant is

(continued)

completely dismantled, the property will be released for unrestricted use. A number of research and demonstration reactors in the United States are now in safe storage. The 10-megawatt experimental sodium reactor in Santa Susana, California, was SAFSTORed when it closed in 1964. Subsequent decommissioning began in 1974 and was completed in 1983. Commercial power reactors in SAFSTOR include LaCrosse Nuclear Generating Station in Wisconsin, put in SAFSTOR in 1988 for a period of 25 years, and Rancho Seco in California, closed in 1989 and now being prepared for SAFSTOR for 20 years.

Entombment (ENTOMB). In this approach, the liquid waste, fuel, and surface contamination are removed to the greatest extent possible soon after shutdown and then the reactor is sealed with concrete or steel. The structure remains entombed for a period of time sufficient to permit the radioactivity to decay to unrestricted release levels. The property must be guarded to protect against intrusion. The NRC will allow this alternative to be used only for nuclear facilities contaminated with relatively short-lived radionuclides, so that all contaminants would decay to levels permissible for unrestricted use within approximately 100 years. Three small research or experimental reactors—in Hallam, Nebraska; Piqua, Ohio; and Ricon, Puerto Rico—have been entombed.

PAST LOW-LEVEL WASTE MANAGEMENT

Particularly during the 1950s, some of this nation's low-level waste was disposed of at sea; however, most has been disposed of in shallow land burial sites. Typically, this waste is placed in containers, separated according to the type of packaging required and the degree of hazard posed, and then placed in trenches. A typical trench is about 600 hundred feet long, 60 feet wide, and 25 or more feet deep. When a trench is full, it is covered with a clay cap or similar low-permeability cover and contoured to control drainage and erosion.

The first commercial low-level waste disposal site opened in 1962 at Beatty, Nevada. By 1971, six commercial sites were operating, but by 1978 three sites—in West Valley, New York; Maxey Flats, Kentucky; and Sheffield, Illinois—had closed. The West Valley low-level waste disposal site, part of a larger Nuclear Fuel Services facility, closed in 1975 when accumulated water overflowed from two of the site's 14 burial trenches. When state authorities in Kentucky discovered that some radioactive material had migrated from the burial site at Maxey Flats, they put such a high surcharge on waste buried there that the operation soon became uneconomical, and the site closed in 1977. The Sheffield, Illinois, site closed in 1978 when it reached its licensed capacity and could not expand because of state opposition and the NRC's denial of permission to open a new trench. All three of these closed sites require extensive cleanup. EPA placed the Maxey Flats site on the Superfund National Priorities List in 1986.

When the first commercial low-level waste disposal sites were selected, there were no uniform regulations for site selection and operation, and technical methods for studying sites were relatively unsophisticated. Some hydrogeologic problems encountered at those sites, including erosion, accumulation of water in trenches, and unexpected complexity in groundwater movement, are attributed to the inadequate earth-science criteria used to select the sites and design the facilities.

Low-level waste facility designers say they have learned a great deal from past difficulties and maintain that problems due to faulty construction and improper siting procedures can be prevented in the future. NRC regulations are now in place to guide the siting, operation, and closure of low-level waste disposal facilities, although EPA had not established environmental standards as of the end of 1992.

In 1992, the low-level waste facilities at Barnwell, South

FIGURE 14. Photo of shallow land burial of low-level waste at Barnwell, South Carolina. ***Source***: Chem-Nuclear Systems, Inc.

Carolina, Richland, Washington, and Beatty, Nevada, were the disposal sites for most types of commercially generated low-level waste. These facilities now have good operating records in meeting current regulations. In particular, by adopting special controls for water management and incorporating natural and engineered systems, the facility at Barnwell, South Carolina, has compiled a successful record of waste containment in a humid environment.

POLICIES AND PROGRAMS

During the late 1970s, the three states with operating commercial low-level waste disposal sites encountered both operational and political problems, and they pressured Congress to pass legislation to relieve them of the complete burden of low-level waste disposal. The governors of Nevada, South Carolina, and Washington demanded improved federal packaging requirements and enforcement. The federal government took no action, and in late 1979, Nevada and Washington closed their sites temporarily because trucks were delivering damaged and leaking nuclear waste containers. At the same time, South Carolina announced that over the next two years the state would reduce by half the amount of low-level waste it would accept annually. The Barnwell, South Carolina, site, the only commercial site located in the eastern half of the country—where the majority of nuclear plants and other waste generators are located—received about 85 percent of all commercial low-level waste disposed of during the 1970s. The states' actions caused alarm among those who used the sites, particularly medical and research facilities with limited storage capacity, and provided some of the impetus behind the passage of the Low-Level Radioactive Waste Policy Act in 1980.

Legislation

With the strong backing of the National Governors' Association, the National Conference of State Legislatures, and most state governments, Congress passed the Low-Level Radioactive Waste Policy Act in December 1980. The act established two major national policies:

- Each state is responsible for assuring adequate disposal capacity for the low-level waste generated within its own borders, with the exception of waste generated by federal weapons or research and development activities.
- The required disposal facilities can best be provided through regional groupings of states allied through interstate agreements called compacts.

Congress must approve a compact ratified by a group of states before it takes full effect. Every five years, Congress may reconsider and withdraw approval of a compact.

Although the law did not require that states work together, Congress added an incentive to encourage them to do so, stipulating that any regional compact could include a provision to exclude waste from outside the region's borders after January 1, 1986.

However, it took much longer than Congress had anticipated for states to reach agreements on compacts, decide on host states, and begin to site facilities. By 1984 it was clear that no new disposal capacity would be available before the 1986 deadline. Nevertheless, citizens in Nevada, South Carolina, and Washington insisted that Congress honor the intent of the Low-Level Radioactive Waste Policy Act and allow them to exclude out-of-region waste from their disposal sites in 1986. Many generators of waste feared they would be left without any way to dispose of their waste. To break this impasse, Congress passed the Low-Level Radioactive Waste Policy Amendments Act in December 1985. The act attached penalties to provisions of the 1980 act; in exchange, the sited states agreed to remain open to the rest of the nation for an additional seven years.

The 1985 Amendments Act provided:

- ▼ A strict timetable that states and compacts must meet for developing low-level waste disposal facilities.
- ▼ Rewards and penalties associated with the milestones in the timetable. If the states do not comply with the milestones, generators can be assessed a penalty surcharge during a specified grace period. If compliance is not achieved by the end of the grace period, the generators may be denied access to the existing disposal facilities. If a state or compact failed to provide disposal capacity by 1993, generators of low-level waste could require that the state take title to the waste, which the state was required to do by 1996.

The act placed caps on the amount of low-level waste that each of the current three disposal sites is required to take and imposed an escalating surcharge on out-of-region waste. The act authorized the three states containing sites and DOE to enforce the penalties, and it gave the federal government responsibility for disposing of greater than class C waste.

In 1992, New York State won a lawsuit overturning the provision of the 1985 act requiring states without access to an operating disposal facility on January 1, 1996 to take title to and possession of all low-level waste generated within their borders. All other provisions of the act were left in force.

Implementation

Forming regions.

The 1980 act did not designate specific regional groupings, and much of the action after its passage concerned negotiations among states to form compact regions and establish formal compact agreements. By 1992, Congress had approved agreements forming nine compact regions: Appalachian, Central, Central Midwest, Midwest, Northeast, Northwest, Rocky Mountain, Southeast, and Southwest. But the compact arrangements and progress on siting remain fluid.

According to plans at the end of 1992, the facility at Richland, Washington, will continue to provide disposal capacity for the Northwest region and, under contract, for the Rocky Mountain region; the facility at Beatty, Nevada, having operated at a very reduced capacity for years, closed at the end of 1992 because Nevada did not renew its license; and the facility at Barnwell, South Carolina, will remain open to the nation, subject to certain conditions, until July 1994, and to the members of the Southeast Compact until 1996, when North Carolina must assume host state responsibilities for the Southeast Compact. Because they have failed to meet milestones in the act, Michigan, Rhode Island, New Hampshire, Puerto Rico, and the District of Columbia are barred from disposing of waste at Barnwell.

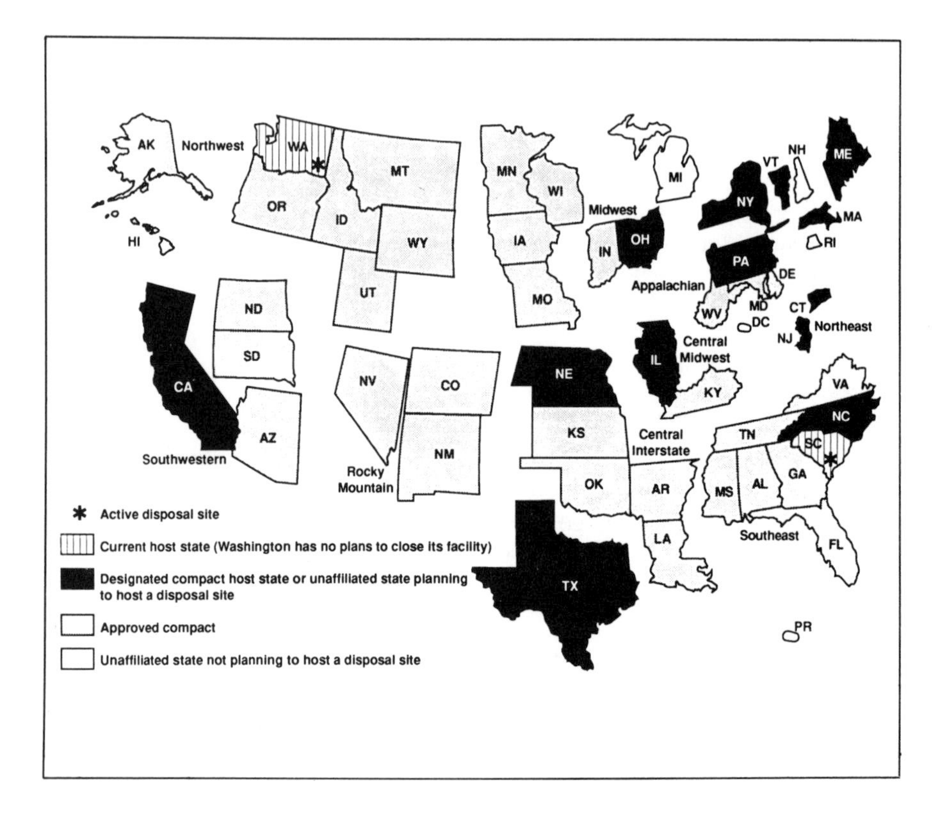

FIGURE 15. Map of Low-Level Radioactive Waste Compact Status as of December 31, 1982. ***Source:*** USDOE, National Low-Level Waste Management Program.

Finding sites.
Before host states can provide the disposal capacity required by the law, they must design a siting process, choose a site and a disposal technology, submit a license application, have it approved, and construct the facility.

States have encountered formidable difficulties in siting new facilities, and by 1992, no state had yet succeeded in licensing the construction of a new low-level waste disposal facility. Three compact host states—California, Illinois, and Nebraska—had license applications under review but each application has encountered difficulties. Ohio, designated as host state for the Midwest Compact, is developing legislation to guide facility siting and development. Connecticut and New Jersey, the only members of the Northeast Interstate Compact, are both responsible for developing disposal capacity, and both are depending on a volunteer process to find a site. North Carolina, host state for the Southeast Compact, is characterizing two sites.

Of the go-it-alone states, Texas has most vigorously pursued a siting process despite a number of setbacks. The authority has purchased and is characterizing a site at Eagle Flat in Hudspeth County.

New York State's siting process reached the point of designating potential sites for study, suffered a major political upheaval, and has begun again. Other go-it-alone states have conducted only low-key siting efforts. Some are negotiating to join an established compact, form a new one with a willing host state, or use capacity at an established region's site.

Building facilities.

New low-level waste disposal facilities must be constructed to meet either the NRC's rule (10 CFR Part 61) or equivalent state requirements in "agreement" (that is, self-regulating) states. The NRC rule includes sections relating to performance objectives, technical requirements, financial assurances, licensing procedures, and state and tribal participation. Performance objectives include provisions aimed at protecting the general population from releases of radioactivity, protecting individuals from inadvertent intrusion into the buried waste, protecting workers during operations, and ensuring the stability of the site after closure. The technical requirements for near-surface disposal include criteria for site suitability, site design, operation, and closure; waste classification and characteristics; and institutional requirements, such as how long the site will have to be guarded after closing.

As a matter of state law or policy, most new facilities will not use shallow-land burial technology. To help states decide what technology to use, the NRC asked the Army Corps of Engineers to conduct a study of alternative methods of low-level waste disposal. These include underground vaults, above-ground vaults, earth-mounded concrete bunkers, mined cavities, and augured holes. The corps study concluded that each of these methods offers some advantage over shallow land burial in meeting the commission's performance objectives. However, the design, construction, and operating costs for each of these alternatives will probably be higher and the operating procedures more complex than for shallow land burial. The corps study cites potential drawbacks, particularly in the use of above-ground alternatives, including increased worker exposure to radiation, complex operational requirements, the need for long-term maintenance, and possible human intrusion.

Since the corps study was conducted, a French low-level waste facility using earth-mounded concrete bunkers has been filled and is being closed. France has built a new fully automated, monitored disposal vault at l'Aube, incorporating the experience gained from operating the first facility. Spain and Sweden have also constructed highly engineered facilities. Most states developing new facilities in this country plan to use engineering technology similar to that used in France and Spain and to draw on the operating experience of the first French facility.

Providing interim storage.

In states where progress is slow and the availability of disposal capacity in the future is uncertain, low-level waste generators are concerned about how and where to store their waste after the Barnwell facility closes to them in June 1994. Licenses for many small generators allow them to store a larger amount of low-level waste than most of them usually keep at the site. Those that have extra space could simply retain more waste at the site than they have in the past. However, some facilities, such as hospitals in crowded urban areas, may not have appropriate unused space.

FIGURE 16. Average yearly low-level waste volume and radioactivity comparison by state and compact, ... which shows the volume of low-level waste and the amount of radioactivity shipped to commercial disposal sites, is based on disposal site records. A six-year average is used because waste volumes and radioactivity can vary considerably year to year. ***Source***: U.S. Department of Energy, National Low-Level Waste Management Program.

Area	*Average Yearly Volume (ft^3)*	*Average Yearly Radioactivity (curies)*
Northeast (NE)		
Connecticut	44,566	71,690
New Jersey	49,773	76,908
	94,339	148,598
Applachian (AP)		
Pennsylvania	140,633	108,371
West Virginia	203	12
Maryland	24,060	4,473
Delaware	1,135	1
	166,031	112,857
Southeast (SE)		
Georgia	38,882	13,903
Florida	29,941	4,811
Tennessee	117,407	1,819
Alabama	36,668	17,159
North Carolina	57,172	20,757
South Carolina	74,695	4,652
Mississippi	11,977	1,560
Virginia	71,998	6,976
	438,740	71,637
Central States (CS)		
Arkansas	10,366	6,014
Louisiana	18,997	2,214
Nebraska	14,301	17,288
Kansas	5,165	1,134
Oklahoma	35,178	55
	84,007	26,705
Midwest (MW)		
Wisconsin	8,308	1,055
Indiana	2,711	100
Iowa	9,288	16,474
Ohio	28,467	2,326
Minnesota	27,941	21,586
Missouri	17,259	1,227
	93,974	42,768
Central Midwest (CM)		
Illnois	153,813	51,868
Kentucky	3,584	257
	157,397	52,125
Rocky Mountain (RM)		
Colorado	8,806	7,867
Nevada	542	136
New Mexico	1,456	26
Wyoming	3	2
	10,807	8,031
Southwest (SW)		
South Dakota	2,102	101
Arizona	23,692	847
California	90,248	15,557
North Dakota	42	21
	116,084	16,526
Northwest (NW)		
Idaho	52	11
Northwest (NW) *(cont.)*		
Washington	42,116	2,822
Oregon	88,210	455
Utah	5,374	42
Alaska	44	258
Hawaii	3,896	5
Montana	91	3
	139,783	3,596
Unaligned		
Rhode Island	744	5
Vermont	6,485	25,435
New Hampshire	180	11
Maine	8,848	1,606
New York	80,416	55,636
Massachusetts	49,074	47,246
Texas	54,759	3,350
Puerto Rico	0	0
DC	897	10
Michigan	23,595	6,137
	224,998	139,436
United States	1,526,161	622,278

Nuclear power plants can store an unlimited amount of low-level waste on site for five years. Some utilities are seeking approval from regulators to build long-term storage facilities. The NRC has been reluctant to endorse long-term storage, fearing that additional storage capacity would relieve pressure on states to site facilities and thus lead to longer delays. However, the NRC is now working on a rule that would allow waste generators to extend the length of time waste is stored on site, if licensees can demonstrate that they have exhausted all other options.

Costs

The per unit cost for disposing of low-level waste has risen dramatically over the past two decades and may increase even more. Past increases are due to surcharges mandated by the 1985 act (no longer applicable after 1992), the reduced volume of waste generated in the United States, and the expense of meeting increasingly stringent operating requirements. Future increases may be caused by rising construction and operating costs of proposed smaller, more highly engineered facilities.

In 1975, the charge for disposing of class A waste was approximately $1 per cubic foot; by 1985, the charge had risen to approximately $20 per cubic foot; and in 1992, base disposal charges averaged about $40 per cubic foot, plus a significant surcharge ranging from $40 to $160 per cubic foot, depending on the disposal facility and how well the state in which the waste was generated had met the milestones set in the 1985 act. Portions of these surcharges were rebated to states and compact regions that had met these milestones and were to be used to develop new disposal facilities.

A large number of individual states and regions propose to build new disposal facilities. If all of these are actually built, they could provide more capacity than is physically needed or economically feasible. However, the overcapacity problem may not arise because some of the plans may not materialize, and because states are scaling the size of proposed facilities to the amount of waste they expect to receive. The greater problem may be the high per unit cost of disposal for these relatively small facilities. Another currently unknown variable is the number of nuclear power reactors that will be closed in the near future and the amount of low-level waste that will be generated by their decommissioning.

As of January 1, 1993, only two low-level disposal facilities were open: in Richland, Washington, and Barnwell, South Carolina. On that date, the base disposal charges at the Richland facility (now open only to members of the Northwestern and Rocky Mountain Compacts) was $36 per cubic foot plus an additional surcharge of $6.50 per cubic foot. A 1991 state law mandates that funds from the surcharge be divided between Benton County (where Richland is located) and the Washington State Economic Development Foundation. To support its activities, the Northwest Compact Commission also charges an annual permit fee to waste generators who use the Richland site. The fee is assessed on a sliding scale based on the volume of waste a generator disposed of at the facility. The projected revenue from permit fees for 1993 is $200,000.

On January 1, 1993, the base disposal charge at the Barnwell, South Carolina, facility, was $59 per cubic foot. Generators of waste from outside the Southeast Compact region are assessed a surcharge of $220 per cubic foot. Under the South Carolina law that allows Barnwell to remain open to out-of-region waste until June 1994, at least $160 of the $220 per cubic foot surcharge must go to South Carolina. The remaining $60 is a contingency fee for the Southeast Compact Commission, to assist in the development of a disposal facility in North Carolina, to support the activities of the Southeast Compact Commission, to handle any legal claims against the Compact, and to compensate South Carolina if South Carolina's $160 share of the surcharge does not cover the state's revenue projections.

Generators within the region are assessed a quarterly regional access fee averaging about $35 per cubic foot. The regional access fees are expected to generate a total of $12 million per year and are to be used to develop a North Carolina disposal facility.

TRANSPORTATION AND LIABILITY

Transportation of radioactive material and liability for nuclear accidents raise some practical concerns: How should nuclear material be packaged to make its transport safe? How will trucks or trains carrying nuclear material be supervised and routed? How will people be compensated if something goes wrong, either during transport or at some other point in the management of nuclear material?

TRANSPORTING NUCLEAR WASTE

Since the beginning of this country's nuclear program, there have been more than 2,500 shipments of spent fuel and many more shipments of low-level waste. The safety record to date is very good. However, shipments of radioactive waste will increase dramatically when any of DOE's major radioactive waste management facilities—a repository for high-level waste, a central storage facility for spent fuel, or the Waste Isolation Pilot Plant (WIPP) for defense transuranic waste—begin operation.

Regulatory and Shipping Responsibilities

Two federal agencies—the Department of Transportation and the Nuclear Regulatory Commission—share responsibility for developing and enforcing safety standards to ensure safe transport of radioactive wastes. To avoid conflicts and overlaps, DOT and NRC have worked out agreements about their respective responsibilities.

Under the Hazardous Materials Transportation Act of 1975 as amended in 1990, DOT has the authority to establish standards on "any safety aspect" of the transport of hazardous (including radioactive) materials "by any mode" in interstate and foreign commerce. DOT sets standards for packaging and shipping for certain low-level radioactive materials; sets requirements for general labeling, handling, placarding, loading, and unloading; and regulates carrier personnel qualifications and shipment routing. DOT is in the process of amending its regulations for domestic radioactive materials transportation to reflect standards developed by the International Atomic Energy Agency in 1985.

The NRC, under the Atomic Energy Act of 1954, has authority to regulate "the receipt, possession, use and transfer of radioactive materials." The NRC sets packaging standards and regulates the shipment and containment security of: certain radioactive materials, including large quantities; special nuclear materials; and spent nuclear fuel shipments to and from commercial nuclear power plants. NRC's standards are consistent with international guidance on the transportation regulations developed by the International Atomic Energy Agency.

Under the provisions of the Nuclear Waste Policy Act of 1982, DOE's Office of Civilian Radioactive Waste Management is responsible for all shipments of high-level nuclear waste and commercial spent fuel to federal facilities. These shipments must comply with DOT regulations. Under provisions of the Nuclear Waste Policy Amendments Act of 1987, DOE must transport commercial spent fuel and high-level waste in NRC-certified shipping casks. State governors or their designees must be notified before shipments of spent fuel are transported through their states. DOE also must provide funds to help states and Indian tribes train local public safety officials to monitor routine transportation and to respond to emergencies.

DOE is required to enter into contracts with producers of commercial spent fuel to take title to the waste when it leaves the commercial site for shipment to a federal repository. Initially signed in 1983, the contracts include provisions that cover transportation from the reactor to the repository or to a federally owned and operated interim storage facility. All costs for the transport of commercial spent fuel are to be borne by the users of nuclear generated electricity through payments to the Nuclear Waste Fund.

Transportation of nuclear waste is of particular concern in states and tribes along the main transportation routes to the proposed nuclear waste disposal sites at the Waste Isolation Pilot Plant near Carlsbad, New Mexico, and Yucca Mountain, Nevada. In the corridor states for WIPP—Colorado, Idaho, New Mexico, Oregon, Utah, Washington, and Wyoming—DOE, state and tribal governments, and several national and regional transportation organizations are actively preparing for potential shipments of transuranic waste to the facility. These preparations include the development of policy and procedures for preventing accidents, responding to emergencies and bad road and weather conditions, conducting inspections, and providing equipment. Although the requirements of the Nuclear Waste Policy Act do not apply to these shipments, DOE has formally agreed to use NRC-certified casks to transport transuranic wastes to the New Mexico facility.

State, tribal, and local governments want to participate in decisions on the content, timing, and packaging of shipments. As of 1992, 45 states and the District of Columbia had passed specific statutes governing radioactive waste transportation, prompted by public health and safety concerns and the perception that federal regulations and enforcement are inadequate. Virtually all states and about 70 localities have issued regulations that address hazardous and radioactive material and waste transportation through permit and notification requirements, routing restrictions, reporting of spills, training of personnel, inspection and enforcement, and additional insurance coverage. DOT has challenged some of these state regulations as conflicting with federal law.

Packaging

Packaging designed for transportation of nuclear material provides the primary barrier to the release of radioactive contents during shipment. DOT and NRC packaging and containment standards for radioactive materials are based on (1) the degree of hazard posed by specific radionuclides to be shipped, (2) the quantity of radionuclides, since greater quantities require more protective packaging, and (3) the form of the radioactive materials. All nuclear waste must be in solid form before being transported. Current DOT and NRC regulations specify the following types of packaging:

> **Strong tight containers** must be highly durable, have tight seals, and act as shields to prevent exposure to handlers and drivers.
>
> **Type A packages** must meet the requirements for strong tight containers and must be capable of preventing spills and leaks under normal driving conditions. Most low-level radioactive waste is shipped by truck in type A and strong tight containers.
>
> **Type B packages** are designed for radioactive materials with a higher level of radioactivity. They must meet all Type A standards and must be able to withstand a severe accident with no loss of shielding and no release of radioactive materials.
>
> **Special shipping casks for spent fuel** are elaborate and rugged forms of Type B packages. Solidified high-level

waste will be shipped in similar heavily shielded casks. These casks generally consist of a stainless steel cylinder inside a heavy metal shield, enclosed in a steel shell. The casks are designed to withstand a sequence of hypothetical tests that encompass a range of very severe accident conditions—including impact, puncture, fire, and immersion in water—without releasing more than a specified small amount of radioactive material. Type B packages for shipping contact-handled transuranic waste to WIPP are also subject to these requirements. It should be noted that analytical methods, rather than actual field tests on sample casks, are normally used to assess the ability of a cask design to pass these tests. However, scale model testing has been used to verify computer model projections and confirm the adequacy of regulatory standards.

FIGURE 17. Photo of truck transporting a spent fuel cask. ***Source:*** USDOE.

To date, most accidents and leakages in transit have involved low-level waste, and no deaths or serious injuries have resulted. In fact, compared to transport of other hazardous materials, radioactive shipments have an excellent record. But questions remain about packaging requirements and the standards for testing. Critics point out that some accidents involve higher speeds than those tested and that some actual fires may have higher temperatures and last longer than was assumed in the tests. The Congressional Office of Technology Assessment (OTA) in its 1985 report, *Managing the Nation's Commercial High-Level Radioactive Waste*, noted the arguments that a drop test of 30 feet onto an unyielding surface and a fire of 1,475 degrees were inadequate to simulate some severe accidents, but commented that some aspects of real accidents might impose less demanding conditions on casks than do the tests. For example, "objects in the real world are not completely unyielding; if struck by a transportation cask, they would absorb some of the energy of the cask." In addition, actual fire is unlikely to surround completely all surfaces of a cask, as specified in the regulatory test.

Some states advocate the full-scale testing of transport casks, but such a requirement was deleted in the compromise that led to the enactment of the Hazardous Materials Transportation Uniform Safety Act in 1990.

Routing

The route of a radioactive material shipment depends on the type of material in the shipment, its size, the distance it must travel, and federal, state, tribal, and local regulations.

Under the Hazardous Materials Transportation Act of 1975 (HMTA), DOT issued two sets of routing regulations in 1981 for highway carriers of radioactive materials. First, a general set of regulations governs radioactive shipments of radiopharmaceuticals, radionuclides for industrial use, and low-level waste which, if properly packaged, are considered to present relatively minimal risks. These regulations allow carriers to use their own discretion in selecting routes. The second set of routing rules applies to large quantities of radioactive materials and is more stringent. Carriers are required to use interstate highways as preferred routes, to avoid travel during rush hours, and to avoid local hazards such as roads and bridges under construction or repair. These rules also permit states to designate alternative preferred routes. Furthermore, drivers must have special training certification and must be notified that they are carrying radioactive materials.

Despite DOT regulations, many states and localities complained that the act and its implementation were flawed by weak federal enforcement, lack of information about hazardous cargoes, inadequate funding for emergency response, and state and local inability to restrict hazardous traffic to safer routes. At least 11 states and many local governments have established their own rules, specifying prenotification requirements, time-of-day restrictions, routes, and special equipment. Subsequent DOT inconsistency rulings and court rulings on law suits brought by DOE and regulated carriers have upheld the agencies' position that federal rules take precedence. Although state permit regulations and other rules that comply with federal laws have been allowed, absolute shipment bans have been struck down unless they can be justified as a safety measure. But the controversy continues.

In 1990, Congress passed the Hazardous Materials Transportation Uniform Safety Act (HMTUSA), which authorized DOT to establish routing guidelines for all hazardous materials (although states can establish routes using federal standards), issue regulations for the inspection of large quantities of nuclear material, and conduct a study to determine whether dedicated trains are the safest method of transporting spent fuel and high-level waste. DOT also is required to examine the factors that shippers and carriers should consider in selecting modes and routes for transporting nuclear waste, and to establish training and grant programs to help prepare local officials to respond to emergencies.

LIABILITY COVERAGE FOR ACCIDENTS

Compensation for a nuclear accident—whether it occurs at a nuclear power plant, a DOE facility, or along a transportation route—is provided for by the Price-Anderson Act. This amendment to the 1954 Atomic Energy Act, first passed in 1957, had two purposes: to insure compensation to the public in the case of a nuclear accident, and to protect the nuclear industry from a potential accident liability so large that it would threaten the future of nuclear power. Renewed in 1966, 1975, and 1988, the act now provides coverage for accidents involving commercial nuclear power plants, certain smaller reactors (such as university research reactors), and nuclear research, fuel processing, waste management, and weapons production activities performed by Department of Energy contractors.

The Price-Anderson Act sets up a two-tier system of insurance against nuclear accidents and a "no-fault" liability system for large accidents. A certain amount of money would be immediately available to compensate individuals for personal injuries or property damage caused by a nuclear accident. If claims exceed the liability limits, Congress would enact legislation to provide full and prompt compensation to the public.

Civilian Activities

The Price-Anderson Act establishes a ceiling on the amount for which a company is liable in a single nuclear accident. A company buys the first layer of insurance (set in 1988 at $200 million for each large reactor site) from private insurance firms. The second layer applies only to operators of large licensed power reactors. If a nuclear accident causes damages exceeding $200 million, each licensed nuclear power plant would be assessed a prorated share of the damages in excess of $200 million, up to $66 million per reactor per accident. Increasing the secondary or potential "deferred premiums" payments that plant owners must make from

$5 million up to $66 million was the major change to the Price-Anderson Act in the 1988 amendments. The NRC must adjust the limit of the deferred premium payment for inflation, not less than once every five years. The deferred premium insurance totaled $8 billion for the 116 commercial reactors covered by the Price-Anderson system as of January 1993.

In exchange for limiting liability to this extent, the Price-Anderson Act in effect imposes "strict liability" on the utility involved in an accident determined by the NRC to be an "extraordinary nuclear occurrence" (ENO). Strict liability means the utility must waive legal defenses against paying claims (up to the ceiling), relieving victims from the necessity of proving that the utility was negligent or "at fault." To recover damages under this provision, affected citizens need to show that their losses were caused by the extraordinary nuclear occurrence, and they must show the amount of their damages.

Department of Energy Activities

The 1988 Price-Anderson Amendments Act sets the liability limit for DOE's contractors' activities at the same amount as the total liability of the nuclear utility industry, currently over $7 billion, with coverage adjusted for inflation not less than once every five years. The act requires the department to indemnify all of its contractors to the full extent of this liability. As with civilian accidents, if damages exceed the statutory limit, the President must send Congress a plan to provide prompt and full compensation for all valid claims. Claims resulting from activities funded by the Nuclear Waste Fund would be paid from the Nuclear Waste Fund. Other claims covered by government indemnity would be paid from general revenues.

Nuclear waste activities are among those for which DOE must indemnify its contractors. Nuclear waste activities are defined in the act as "involving the storage, handling, transportation, treatment, or disposal of, or research and development on, spent nuclear fuel, high-level radioactive waste, or transuranic waste." The act also adds "a nuclear incident [that] . . . arises out of, results from, or occurs in the course of, nuclear waste activities" to the list of situations that could qualify as an ENO or extraordinary nuclear occurrence. The act also identifies specific activities at the Waste Isolation Pilot Plant covered under the act.

Transportation

In the event of an accident involving the transportation of nuclear waste, the source of the funds for public compensation depends on whether the particular shipment falls under the coverage mandated for NRC licensees, such as commercial nuclear power plants, or under DOE indemnification of its contractors.

A shipping accident involving NRC licensees would be treated like an accident at a nuclear power plant, with a total potential compensation of over $7 billion. These provisions cover all shipments of nuclear material to or from nuclear power plants, including enriched fuel sent to a power plant and low-level waste shipped to a disposal site.

Shipping accidents involving companies operating under contracts with DOE are covered by indemnity from the federal government. DOE, in effect, executes an indemnity agreement with the contractor to cover damages up to the liability limit. Shipments in this category include spent fuel transported to a storage or disposal facility; high-level waste transported from one storage facility to another, or to a repository such as the one proposed for Yucca Mountain; spent fuel or low-level waste from government, research, or foreign reactors en route to storage, to a repository, or to disposal facilities; enriched uranium hexafluoride shipped from an enrichment plant to a fuel fabrication plant; nonirradiated fuel shipped from a fuel fabrication facility to a federal government or research reactor; and uranium hexafluoride transported from a conversion facility to an enrichment plant.

If the NRC or DOE determines that a transportation accident is an "extraordinary nuclear occurrence," then the "strict liability" provision described above would come into play. Thus far, no power plant or transportation accident has been so designated. In the absence of an ENO, liability for an accident in a nuclear waste activity covered by the Price-Anderson Act is generally determined under state law.

DEFENSE WASTE

For more than 40 years, the U.S. government has designed, produced, maintained, and dismantled nuclear weapons in a huge industrial complex of factories and laboratories built mainly in the 1940s and 1950s. Almost all of these aging facilities and their sites are now contaminated with radioactive and other hazardous wastes. In 1989, DOE began developing plans to make the weapons complex smaller and more efficient by closing some facilities, consolidating activities at a few facilities, and modernizing those few. The department also initiated a 30-year program to alleviate the health, safety, and environmental hazards resulting from years of weapons production.

SOURCES OF DEFENSE WASTE

In 1991, the Bush administration decided to dismantle a significant number of existing weapons and perhaps assist the former Soviet Union in dismantling its nuclear weapons. In 1992, the U.S. government decided to cease producing new nuclear warheads.

These changes mean that the major sources of nuclear waste in the weapons complex will be in the cleaning up of contaminated sites, dismantling of weapons, and decommissioning of facilities, rather than in the production and testing of weapons. The task is immense. DOE estimated in 1992 that it must clean up and manage the waste at more than 100 contaminated installations in 36 states and territories.

Cleaning up sites. Widespread environmental contamination is present at almost all weapons research, production, and testing sites. According to a 1991 report of the Office of Technology Assessment (OTA), "There is evidence that air, groundwater, surface water, sediments and soil, as well as vegetation and wildlife, have been contaminated at most, if not all, DOE nuclear weapons sites." Most of this contamination occurred as the byproduct of one or more stages in the nuclear fuel cycle that supported the manufacture of nuclear weapons, from uranium mining through spent fuel processing and weapons manufacture. The testing of nuclear weapons also has contaminated additional sites, most notably the Nevada Test Site. Underground test explosions leave large amounts of buried waste that may be irretrievable and will require long-term surveillance. The operation of test facilities themselves has also generated hazardous, radioactive, and mixed waste that must be stored and disposed of.

The severity and types of contamination at these sites vary greatly, and so will the waste created by the cleanup. The contamination includes large quantities of contaminated soil; high-level radioactive waste stored in ways that threaten to harm humans and the environment; stored and buried transuranic waste; buried low-level radioactive waste; and buried and stored mixed waste. At many sites the contamination, consisting largely of hazardous organic solvents, is moving beyond the boundaries of the facilities.

Dismantling weapons. Nuclear weapons are disassembled into their various components at the Pantex Plant near Amarillo, Texas, where they were assembled. The major weapons components, which include radioactive materials such as enriched uranium, plutonium, tritium, and depleted uranium, could be either reused, stored for possible future use, or treated as waste. The process of dismantling weapons produces waste similar to that generated when the weapons were manufactured: transuranic waste, low-level waste, and mixed waste. If some of the recovered nuclear material is to be discarded, it will be classified as high-level waste.

Decommissioning and decontaminating facilities. At the end of 1991, DOE listed 500 contaminated facilities—such as reactors, processing plants, and storage tanks—to be cleaned and managed by the department's decommissioning and decontamination (D&D) program. D&D will produce large amounts of building material and rubble and varied kinds of waste—high-level, transuranic, low-level, and mixed.

As DOE reduces and reconfigures the weapons complex, the D&D program will grow. The technology and cost required to complete these tasks are not yet completely known. The department also is responsible for decommissioning facilities contaminated by the Manhattan Project, which developed the atomic bomb under the Formerly Utilized Sites Remedial Action Program, and for stabilizing or otherwise managing uranium mill tailings at inactive sites under the Uranium Mill Tailings Radiation Control Act.

OPENING UP THE WEAPONS COMPLEX

From the beginning of the federal government's nuclear weapons research and production program in 1944 until the late 1980s, nuclear weapons facilities were largely closed to outside inspection or regulation. The government's paramount objectives throughout the weapons complex were weapons production and maintenance of secrecy. Waste management and environmental protection were given much less consideration. The Atomic

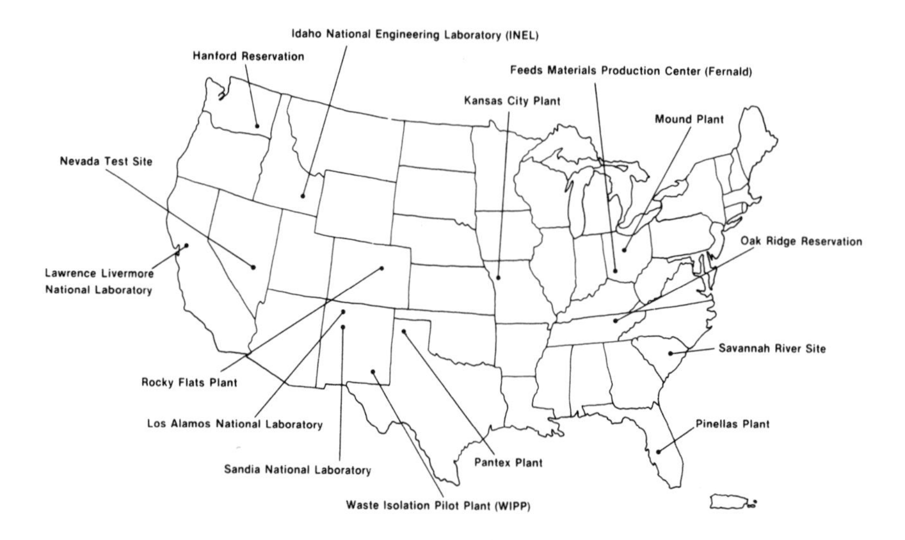

FIGURE 18. Department of Energy Weapons Complex. ***Source:*** Office of Technology Assessment, *Complex Cleanup: The Environmental Legacy of Nuclear Weapons Production*, 1991.

Energy Commission and its successor agencies, the former Energy Research and Development Agency and the current Department of Energy, set their own standards and regulated their own actions.

Even during the early years, however, some people within the weapons complex raised questions about waste disposal practices. In 1948, the Atomic Energy Commission's Advisory Committee on Nuclear Safety pointed out that long-term contamination would result from the continuation of current disposal practices at Hanford, such as storing high-level waste in single-wall tanks and discharging liquid waste to seepage ponds and waste pits to percolate through the soil. The committee recommended such practices be discontinued, but its recommendations were rejected, in part because the AEC said it lacked the money to use other disposal methods.

During the 1970s and 1980s, public awareness of environmental issues increased dramatically, and public acceptance of government sovereignty and secrecy decreased. Congress and many states passed laws to protect public health and the environment from the actions of local, state, and federal governments, as well as those of private interests. However, DOE maintained that, for national security reasons, these laws did not apply to the department's activities.

In 1984, the Legal Environmental Assistance Foundation (LEAF) in Knoxville, Tennessee, and the Natural Resources Defense Council brought suit (*LEAF* v *Hodel*) against the Department of Energy, challenging the department's right to set and enforce regulations for itself. The court found that the department was in violation of the Clean Water Act and the Resource Conservation and Recovery Act at the Y-12 Plant on the Oak Ridge Reservation. As a result of this ruling, the federal government acknowledged that federal environmental laws, as well as certain state and local laws, apply to weapons production activities. This change in policy has slowly opened up the weapons complex to external review and led to significant legal obligations to comply with environmental standards.

In 1989, Secretary of Energy James Watkins announced that compliance and cleanup would now have higher priority than weapons production, and that the department's structure was being reorganized to place responsibility for cleanup and waste management under one manager in the newly created Office of Environmental Restoration and Waste Management.

The new office faces perplexing problems. Throughout the weapons complex, sites are seriously contaminated and facilities are in violation of environmental regulations. Currently there are not enough disposal facilities for the waste that would be generated by cleaning up those sites and facilities as well as by any future weapons production or testing. The extent and nature of risk posed to health and the environment are not yet fully understood. The technology needed to solve some identified problems is not yet developed. Laws and policy goals are complex, overlapping, and potentially conflicting.

FACILITY CLEANUP AND COMPLIANCE

DOE must clean up more than 8,700 contaminated sites located throughout the United States, including 5,000 sites now managed under the Uranium Mill Tailings Remedial Action Program (UMTRAP) and the Formerly Utilized Sites Remedial Action Program (FUSRAP). The department must also bring its current operations into compliance with environmental regulations and other legal requirements.

Planning

Details about environmental cleanup and waste management are worked out in part during DOE's five-year planning process and through the negotiation of agreements between the department and federal and state regulators. The Programmatic Environmental Impact Statement for Environmental Restoration and Waste Management (PEIS) now being written, will significantly shape overall approaches and long-term management policies.

Five-year planning process. First published in August 1989 and updated each year, the Five-Year Plan describes the specific actions that DOE plans to take to achieve compliance with environmental laws, to clean up the weapons complex, to manage weapons complex waste, and to develop new technical methods for these tasks. It also discusses DOE's goals, its planning assumptions and financial and workforce needs, the regulatory requirements behind proposed actions, and progress in achieving compliance. The State and Tribal Government Working Group (STGWG) reviews the plan annually. STGWG's members include representatives of 17 states, five Indian tribes, three national organizations of state officials, and the federal Office of Management and Budget. DOE also holds a "stakeholders forum"—a meeting to which the department invites people representing diverse points of view and expertise—to discuss each draft revision. The department also asks for public comment.

A section of the 1991 National Defense Authorization Act (PL 102-190) wrote into law the requirement for and the description of a five-year plan to be issued annually, a schedule for its development, and the requirement for its review by "the Governors and Attorneys General of affected States, appropriate representatives of affected Indian Tribes, and the public."

DOE manages the weapons complex through field offices. The field offices manage the private companies which, under contract with DOE, carry out the day-to-day activities at that field office's sites. Each field office prepares and annually updates site-specific plans. These plans list the actions the field office will conduct at each site in order to meet five-year plan goals and commitments. The site-specific plans are released for public review.

Agreements. Federal environmental laws require DOE to enter into several kinds of legally enforceable compliance agreements with the EPA and state regulators, establishing procedures and schedules for meeting regulatory requirements. These include Federal Facilities Agreements, Federal Facility Compliance Agreements, Settlement Agreements, and Consent Orders. In particular, at sites on the National Priorities List (Superfund), DOE must enter into interagency agreements with EPA. EPA encourages affected states to join as parties to these agreements.

By mid-1992, DOE and EPA signed a total of 87 compliance and cleanup agreements, with 27 still being negotiated. Many of these agreements also involve the states. Approximately 20 agreements have been revoked because violations have been corrected or because the activities covered have been incorporated into more comprehensive, integrated agreements.

DOE and states also may negotiate Agreements-In-Principle (AIP), in which DOE agrees to provide a state with the access and the funding needed for monitoring air, groundwater, and surface water around a site and for overseeing the department's compliance with environmental laws and regulations.

Indian tribes also have very specific authority, responsibilities, and rights that must be taken into account by DOE in determining how to clean up sites and manage waste (see Chapter 3: Radioactive Waste Issues in Indian Country).

Programmatic environmental impact statement. Another significant project that will shape more long-term planning for cleanup and compliance is the Programmatic Environmental Impact Statement for Environmental Restoration and Waste Management (PEIS). This document, which is required by the National Environmental Policy Act (NEPA), will assess and compare the potential environmental impacts of alternative general cleanup and management approaches. For example, the department will choose whether, in general, its approach will be to leave contaminants in place and either treat or contain them where they lie, or to move them to another place and either treat or contain them there. The purpose of the PEIS is to insure that, in choosing among alternative approaches, DOE considers potential environmental impacts as well as such factors as costs, public health risks, and risks to workers. DOE plans to release a draft of the PEIS in 1993 for public comment.

After the PEIS is prepared, DOE will complete a study setting out a long-range strategic plan for developing alternative waste management systems described in the PEIS. The department is also preparing a separate programmatic environmental impact statement on the proposed reconfiguration of the weapons complex.

Policy Issues

The 1976 Resource Conservation and Recovery Act and the 1980 Comprehensive Environmental Response, Compensation, and Liability Act established three general policies governing nuclear weapons complex management and cleanup of hazardous waste:

- ▼ Reduce or eliminate the amount of hazardous waste being produced.

MAJOR ENVIRONMENTAL LAWS

RCRA. The 1976 Resource Conservation and Recovery Act was enacted to address the widespread problem of contamination from the disposal of municipal and industrial hazardous and solid waste. The act established an important national policy: governments and private companies must reduce or eliminate the amount of hazardous waste they produce and must treat, store, or dispose of existing waste in ways that minimize its threat to health, safety, and the environment. The program is managed by the U. S. Environmental Protection Agency (EPA) or EPA-authorized states.

RCRA requirements are enforced through a permitting process. That is, in order to operate a facility to treat, store, or dispose of hazardous or mixed waste or a plant that generates such waste, the Department of Energy or other facility operators must have permits from EPA or EPA-authorized state regulators, or must qualify for interim status, which allows a facility to continue to operate while its operator seeks a permit. Before a permit is granted, an applicant must report about the facility's procedures and must take corrective action to prevent or otherwise address the release of hazardous material into the environment. Most weapons complex facilities are operating under interim RCRA status.

Under the Hazardous and Solid Waste Amendments of 1984 (HSWA) DOE is required to eliminate contaminant releases at or from its RCRA facilities according to a schedule specified by EPA. By 1992, DOE was carrying out corrective action at most weapons sites and planning to comply fully with the act by 1997. The Federal Facilities Compliance Act requires compliance by October 1995.

(continued)

CERCLA/Superfund. The Comprehensive Environmental Response, Compensation, and Liability Act of 1980 (also known as Superfund) sets another important national policy: governments and private interests must clean up hazardous and radioactive substances that could endanger public health, welfare, and the environment. The act provides the U.S. Environmental Protection Agency with the authority to assess contaminant releases from abandoned waste sites, categorize sites according to their risks, and include them in the National Priorities List if the agency considers their cleanup a national priority. This authority covers both radioactive and hazardous contaminants. In addition, the Superfund Amendments and Reauthorization Act (SARA) of 1986 explicitly includes federal facilities, such as those in the weapons complex, under CERCLA.

After a site is placed on the National Priorities List, EPA, often in partnership with states and with input from the public, closely oversees the processes and the schedules for studying and cleaning the sites. DOE must investigate and report about the contamination at those facilities according to set schedules, and it must enter into interagency agreements with EPA concerning cleanup schedules, operation and maintenance of sites, and cleanup options considered and chosen. States also may be parties to the agreements. As of 1992, eight weapons facilities were on the National Priorities List authorized by CERCLA.

FFCA. The Federal Facilities Compliance Act of 1992 allows the U.S. Environmental Protection Agency (EPA), EPA-authorized states, and courts to impose on DOE facilities the full range of punitive sanctions available under RCRA or state and local hazardous and solid waste laws. The FFCA amends RCRA's federal facilities provision to include an express waiver of immunity from punitive penalties, making federal facilities subject to the same sanctions as any other polluter. The act requires DOE to submit two reports to EPA and to the states in which mixed waste is generated or stored by the department: a state-by-state inventory of mixed wastes and a national inventory of mixed waste treatment capacities and technologies. The act also requires DOE to develop and gain regulatory approval of a treatment plan for each facility that generates or stores mixed waste.

NEPA. The National Environmental Policy Act of 1969 mandates that all federal agencies and departments take into consideration the impacts their actions may have on the environment. The Council on Environmental Quality has been responsible for specifying how federal agencies are to comply with the act, although the Clinton administration has proposed reassigning that responsibility.

NEPA requires that agency actions undergo environmental review early in the planning process and that the review process be open to public participation. This review often results in the preparation of an Environmental Assessment or an Environmental Impact Statement, usually on a specific project or regulation. An EIS prepared for an entire program of agency activities is called a Programmatic EIS (PEIS). The Department of Energy is currently preparing a PEIS for waste management and environmental restoration at the weapons complex and a separate PEIS for consolidating and modernizing weapons production facilities. Major DOE sites have prepared or will prepare site-wide EISs.

CAA. The Clean Air Act authorizes the Environmental Protection Agency to set standards for radioactive and hazardous pollutant emissions into the atmosphere at various facilities, including facilities for weapons production, high-level waste and low-level waste disposal,

(continued)

and uranium mill tailings sites. In 1989, EPA issued emissions standards for weapons facilities, low-level waste disposal facilities, and uranium mill tailings sites. States can set more stringent standards than those required by EPA.

CWA. The Federal Water Pollution Control Act, as amended by the Clean Water Act, prohibits the discharge of any pollutant, including radioactive and hazardous wastes, into any U. S. navigable waters without a permit. Discharge limits in permits are determined in accordance with water quality standards. The CWA also includes provisions for monitoring, recordkeeping, and reporting of discharges. EPA and EPA-authorized states have the authority to issue permits. The CWA expressly requires that all federal facilities comply with CWA standards and permitting requirements.

- ▼ Manage hazardous waste in ways that minimize its threat to health, safety, and the environment.
- ▼ Clean up sites contaminated with hazardous or radioactive substances.

These and other laws and regulations also specified legal means for ensuring that the government acts to carry out these policies, and they established requirements for involving the public in decision-making. Additional defense waste policy issues are still evolving.

Thirty-year goal. In 1989, DOE set a goal of 30 years to clean up and restore the environment at its nuclear sites. Many people familiar with the technical problems consider this timetable unrealistically short, particularly since the nature and extent of contamination is unknown at many sites, and the technology for cleaning up many kinds of contaminated sites is still undeveloped. However, in the section of the 1991 National Defense Authorization Act that mandates a five-year plan, Congress also charges DOE with meeting the goal of cleanup and compliance by 2019.

Land use. The federal government in some agreements has stated that it intends to clean contaminated land sufficiently so that its future use can be unrestricted. Debate over whether unrestricted use is a feasible goal at specific sites and how much money should be used to achieve it are issues to be addressed in the PEIS on environmental restoration and waste management.

EPA's Superfund program began with high expectations for the eventual use of contaminated land, but at some sites EPA has had to issue waivers, acknowledging that it is impossible or technically impracticable to clean them; instead, at least for the time being, the agency will allow those responsible for the site just to contain the contaminants and monitor how well the containment is working. Following this course will mean that some land cannot be returned to unrestricted use.

Priorities. The general goal of achieving cleanup and compliance by 2019 must be broken down into decisions about what to do first and how to do it, both in the overall weapons complex and at each facility. In general, DOE says its policy is to respond immediately to problems that in the department's view pose imminent risks or threats to human health or the environment, to develop better technology and solutions to problems for which there are now no clear answers, and to work with states and other federal agencies to plan priorities for addressing other problems. Details about how and when to clean individual sites or bring facilities into compliance are hammered out in the negotiations of binding agreements with state and federal regulators. The precise means by which DOE will determine its priorities is a source of concern to tribes, state and federal regulators, and the public.

Research. Many issues remain unresolved concerning both the direction of research and whether or not the government does or should have a "bias towards action" rather than toward seeking better solutions or continuing to study a site. Should technology development be focused on the most common contamination problems, or should the sites with more difficult or unusual problems also have the benefit of sophisticated research? Under what conditions is it wise to proceed with cleanup using a technique that might or might not work in the long run? Under what conditions is it wise to wait for a better technology to be developed?

Risks. Other debates concern the validity of different methods for estimating, comparing, and evaluating present and future risks to the public, to workers, and to the environment from contaminated sites and waste management facilities. Various parties also differ in their views on the reliability and adequacy of the data DOE uses to assess risks and its methods for factoring those risks into decisions.

Worker health and safety. As DOE gears up to clean more weapons sites and build more waste facilities, an increasingly large number of people will be working in nuclear and hazardous waste cleanup, including the removal or treatment in place of contaminated soil, the dismantlement of buildings, and the construction and operation of waste treatment and disposal facilities. DOE, states, labor organizations, and the public will need to work together to discern how to establish and run effective worker health and safety programs and how to evaluate risks to workers.

DOE facilities are subject to DOE Orders rather than to the federal Occupational Safety and Health Act (OSHA) regulations. An extensive 1989-92 evaluation of DOE facilities reported numerous failures to comply with DOE Orders on worker health and safety. Most were site or facility-specific, such as unsafe electrical connections or improper labeling of hazardous materials. But others involved a fundamental failure to understand safety procedures and requirements. The department says it has made progress in correcting these situations and is developing and implementing new health and safety requirements.

DOE will need to acquire skilled, experienced personnel to support worker health and safety programs and will need to consult with labor representatives and others to resolve health and safety issues. These include means for ensuring that proper training is provided not only to DOE employees and contractors, but also to subcontractors. Workers and the public also will be interested in defining means for communicating and evaluating worker health and safety risks, and for exploring the relation between risks to workers and risks to the public. For example, how great a risk should workers take today in order to prevent future risks to the public and the environment?

Costs. The budget for DOE's Environmental Restoration and Waste Management program has increased from less than $4 billion in fiscal 1991 to approximately $5.5 billion in fiscal year 1993. This increase resulted partly from expanding the program. For example, by 1993 the Office of Environmental Restoration and Waste Management had taken over responsibility for the large Hanford Reservation in Washington and the Feed Materials Production Center at Fernald, Ohio, and was soon to be responsible for the Rocky Flats Plant in Colorado. In addition, costs have and will continue to increase as the program moves from the early phase of studying problems to actually cleaning up sites and facilities and building waste treatment and disposal facilities. Estimates of the total amount needed to complete the 30-year program have increased to more than $160 billion. As costs increase, so do public and congressional expectations of results and questions about how wisely money is being spent.

CLEANUP AT HANFORD

Hanford is one of the oldest and largest nuclear facilities in the nuclear weapons complex. It is located on a 560-square-mile tract of desert along the Columbia River in southeast Washington. From 1943 until 1988, Hanford produced much of the plutonium used in the U.S. nuclear weapons program. It is now the site of a massive cleanup and waste management effort that is predicted to cost upwards of $57 billion over a 30-year period. The attention given Hanford by DOE and by a concerned public is making it a test case for how this nation will deal with the complicated and expensive task of cleaning up its nuclear weapons complex.

The cleanup is being carried out under a 1989 negotiated compliance agreement (commonly referred to as the Tri-Party Agreement) between DOE, EPA, and the Washington State Department of Ecology. The agreement requires DOE to provide for the safe management and disposal of liquid and solid wastes currently stored at Hanford and to clean up contaminated soil and groundwater at the site. It sets legally binding milestones over a 30-year period to achieve these goals and for bringing Hanford into compliance with CERCLA, RCRA, NEPA, and applicable state environmental and health standards. These milestones may be extended or added to by agreement of all parties. EPA and the state are responsible for monitoring DOE's performance and enforcing the terms of the agreement.

The safe management and disposal of wastes stored at Hanford pose an enormous challenge. Hanford is home to almost two-thirds, by volume, of all of the solid and liquid hazardous and radioactive wastes created by the U.S. nuclear weapons program since 1943. Much of this waste is high-level liquid waste (approximately 65 million gallons) stored in 177 underground tanks. Some of these tanks are presumed to have leaked and others may pose a risk of explosion due to the buildup of hydrogen or ferrocyanide in the tanks. The Tri-Party agreement calls for DOE eventually to remove and process the wastes for permanent disposal.

The containment and clean up of contaminated soil and groundwater at Hanford will be an equally daunting task. DOE has estimated that, since 1943, approximately 440 billion gallons of liquid waste have been disposed of in the ground, primarily at 300 sites within the Hanford Reservation. In addition, DOE estimates that as much as one million gallons of high-level mixed waste has leaked from storage tanks into the ground.

In all, there are approximately 1,100 contaminated soil sites within the four areas of Hanford that were added to EPA's National Priority List of Superfund sites in 1989. Several sites have already been subject to expedited cleanup actions to reduce or eliminate potential environmental threats. At most sites, however, extensive characterization work will need to be done before a remedy is selected and actual cleanup is undertaken.

Some of the hazardous and radioactive wastes in the soil have come into contact with groundwater under Hanford and are migrating; low levels of radionuclides have been detected as far as 200 miles downstream in the Columbia River. It is currently estimated that approximately 100 square miles of groundwater under Hanford have been contaminated with radionuclides of uranium, strontium, cesium, and tritium as well as chemical contaminants of nitrates and carbon tetrachloride.

In addition to the wastes already on site, current waste management activities at Hanford continue to generate both solid waste and liquid effluent. Although some waste reduction has been achieved, approximately 400,000 gallons of liquid waste, produced mostly by

(continued)

cooling and ventilation systems, were expected to be released into the soil in 1992. DOE is in the process of updating and constructing treatment facilities, and the Tri-Party Agreement sets a deadline of 1995 for DOE either to treat or eliminate most of these discharges.

The challenge of cleaning up at Hanford lies not only in the vast amount of waste involved, but also in the lack of proven, economical technologies available to address many of Hanford's problems. Methods for cleaning groundwater do exist, but they are extremely expensive and only remove a portion of the contamination. Moreover, they would have to be continued for at least several generations. Options for restoring the soil include digging it up and washing or burning it, or treating it in place, but again, these methods are expensive and their effectiveness unproven. Consequently, some areas of Hanford will probably not be returned to unrestricted use; contaminants will have to be contained and monitored until technologies are developed to remediate these sites.

The level of risk to public health and the environment posed by the contamination at Hanford is not yet fully understood or agreed upon by DOE and other affected parties. DOE has stated that contamination at weapons complex sites does not pose an immediate or near term health risk to persons living off site. A 1991 report by the Office of Technology Assessment concluded, however, that additional information on potential pathways of past, present, and future human exposures and the biological effects of contaminants is needed in order to assure that weapons complex sites do not pose a health threat to workers or the public. The Hanford Environmental Dose Reconstruction Project is one effort to estimate the doses and consequent health effects experienced by local residents in the 1940s and 1950s, when radioactive iodine from Hanford reactors was periodically released into the air, and during the period from 1964 to 1966, when radionuclides were released directly into the Columbia River.

The Tri-Party Agreement provides for public participation in many decisions related to the clean up as required by NEPA, CERCLA, and RCRA. Additional avenues for public comment are provided for by the Washington Administrative Code and the State Environmental Policy Act (SEPA).

Advisory groups are playing an increasingly important role in decision-making at Hanford. In an innovative approach to resolving conflicts at Hanford, DOE in 1992 contracted with an independent facilitator to structure and carry out a process to develop land use recommendations. The Hanford Future Site Uses Working Group, whose 48 members represented a wide range of governmental, tribal, citizen, and other special interests, recently completed work on its recommendations for alternative future land uses and cleanup options. These recommendations will be incorporated into future decision-making processes, including the Hanford Remedial Action environmental impact statement and the Hanford site planning process. The Nuclear Waste Advisory Council, established by the state of Washington, provides advice to Washington State Department of Ecology on a variety of issues of concern related to Hanford. It is expected that additional site-specific advisory groups will be established at Hanford as the cleanup progresses.

In addition to advisory groups, there are several citizen and environmental groups that actively carry out public information and advocacy at Hanford. They include the Hanford Education Action League, Heart of America Northwest, Northwest Environmental Advocates, and the Nuclear Safety Campaign.

WASTE MANAGEMENT

High-Level Waste

The nuclear weapons program generated the first high-level waste in the mid 1940s, when fuel rods irradiated in reactors were processed to recover the uranium and plutonium needed for weapons production. The resulting liquid high-level radioactive waste was stored in large, single-wall carbon steel underground tanks built for this purpose at the U.S. Hanford Reservation at Richland, Washington. Federal officials assumed permanent disposal would be provided later.

Since the waste was acidic, workers neutralized it with sodium hydroxide before pumping it into the tanks. Unfortunately, this solution to one problem led to two others. Neutralization increased the waste volume and created a sludgelike sediment in the tanks that has complicated subsequent efforts to exhume and solidify the waste.

In the early 1950s, DOE's predecessor agency, the Atomic Energy Commission, built double-wall rather than single-wall tanks to store defense-related high-level waste at the Idaho National Engineering Laboratory (INEL) and the Savannah River Site in South Carolina. At INEL, where the tanks were stainless steel, the waste was initially stored untreated (that is, not neutralized). At Savannah River and later at Hanford, where the double-walled tanks were carbon steel, the acidic high-level waste was neutralized.

In the 1960s, the Atomic Energy Commission began the process of solidifying the stored high-level waste to stabilize it and reduce the volume. The neutralized waste at Hanford and Savannah River was partially evaporated, leaving a reduced mixture of liquid, sludge, and salt cake in the carbon steel tanks. At the Idaho Falls facility, where the high-level waste had not been neutralized, it was converted into calcine, a dry granular material. The calcine is stored in stainless steel bins, and the bins are placed in underground concrete vaults.

In 1956, the first leak from the tanks at the Hanford Reservation was detected. In 1960, about 100 gallons of liquid high-level waste leaked from a Savannah River tank and contaminated the groundwater. In the late 1970s, the federal government began to give serious attention to more effectively isolating waste generated by weapons production. Twenty new tanks with improved leak-detection devices were built at Hanford and 27 at Savannah River. The original intent was to pump the liquid high-level waste from the old tanks to the new ones, but at Hanford a serious obstacle was discovered. A nitric acid process that could redissolve the neutralized sludge and solidified salt cake would also corrode the walls of the old tanks, which remain radioactive. Furthermore, steelwork protruding into the sludge from the floor of the tanks at Savannah River interfered with attempts to remove the sediment mechanically.

DOE reported that, as of December 1992, 67 of the 149 Hanford single-wall tanks were known or presumed to be leaking and to have released a total of approximately 750,000 gallons of liquid waste into the ground. This estimate did not include liquid waste that had been deliberately discharged from tanks into cribs and which eventually percolated into surrounding soil. Nor did it include cooling water that might also have leaked into the ground. In 1991, DOE estimated that more than 50,000 and up to 800,000 gallons of cooling water had leaked from a single tank at Hanford. None of the 28 newer tanks with secondary carbon steel barriers is known to be leaking.

High-level waste from weapons production is now stored at the Idaho National Engineering Laboratory, the Hanford Reservation in Washington, the Savannah River Site in South Carolina, and the West Valley Demonstration Project (formerly the Nuclear Fuel Services Plant) in New York. Managing these wastes poses some difficult problems. Uncertainties remain about the best methods for maintaining tanks safely (particularly at Hanford), removing the tank waste, and preparing high-level wastes for storage and disposal.

Tank safety. The waste in many tanks is very corrosive and, over time, may leak through the tank walls. As noted above, this has occurred in several tanks, particularly in the single-wall tanks at Hanford. In addition, the mixture of wastes in a few of the tanks at Hanford produces flammable hydrogen gas that has the potential to explode. DOE has taken and continues to take actions to reduce the likelihood of fires or explosions in the older single-wall tanks or the newer double-wall tanks at Hanford. Other potential problems with reactive organic compounds are just now being recognized and analyzed. DOE has put approximately 49 of the tanks on a "watch list" for special monitoring and care. In its 1990 report, DOE's Advisory Committee on Nuclear Facility Safety stated that the "Hanford Tanks present a serious situation, if not an imminent hazard."

Waste removal. The presence of sludge at the bottom of the tanks, the lack of specific knowledge of the chemical and radiological contents of some tanks at Hanford, and the cooling coils built in some tanks, especially at the Savannah River Site, make it difficult to remove the waste. The Department of Energy has developed and demonstrated a process for completely emptying waste storage tanks at Savannah River and plans ultimately to empty all high-level waste tanks there.

The department is investigating methods for identifying the characteristics of the wastes at Hanford. Some scientists question whether detailed characterization of each Hanford tank is necessary before waste is removed for treatment, since characterization may expose workers to potentially harmful amounts of radiation, is very expensive, and may not be technically required before proceeding. Other factors that must be weighed are whether it is possible or wise to remove all of the sludge from the tanks, and, if not, whether the waste can be satisfactorily stabilized in place.

Solidification. Many scientists have long considered incorporation into ceramic or glass (vitrification) the best way to immobilize high-level nuclear waste before disposal. Both France and Great Britain operate glass plants that vitrify liquid high-level waste from reprocessing their own or other countries' spent fuel. DOE plans call for operating facilities at Savannah River, Hanford, and West Valley to convert high-level liquid waste into borosilicate glass. DOE may opt to vitrify the high-level waste at the Idaho National Engineering Laboratory into a ceramic rather than glass form.

DOE chose borosilicate glass as the form for the waste at Hanford, Savannah River, and West Valley based on the overall ranking of borosilicate glass in a study comparing alternative waste forms according to such qualities as leach resistance, mechanical strength, stability in the presence of radiation and heat, and ability to handle variations in the composition of the waste being treated. During the process for producing the glass, waste and borosilicate glass-forming material would be fed as a slurry into a glass melter and heated to temperatures above 1,000° C. The resulting molten slurry would be poured into stainless steel canisters to cool and harden into glass. The canisters would be tightly sealed by welding and stored until they are disposed of in a geologic repository.

Plans call for separating the wastes that are removed from the tanks. The highly radioactive portion would be solidified for later disposal in a geologic repository, and the low-activity portion would be disposed of on site, immobilized in concrete or grout. Other difficult-to-move wastes in the Hanford tanks may be stabilized in place. As of 1992, the facility at the Savannah River Site has been completed and is undergoing extensive tests before operations are scheduled to begin in 1994. Ground was broken at Hanford for a vitrification plant, but the plant will not be constructed until the plant at Savannah River has been run successfully and the characteristics of the material to be vitrified (separated liquid high-level waste from the tanks) are better understood. Under a joint federal-state project, the West Valley reprocessing plant has been decontaminated and a ceramic melter installed to process the high-level radioactive waste into a glass form. By December 1991, after some steps in the processing had been carried out, 1,729 cubic meters of concentrated high-level waste remained at West Valley awaiting solidification.

Storage and disposal. The sealed canisters of vitrified waste must be stored until a permanent nuclear waste disposal site is available. The current plans are to dispose of the defense waste in the same repository as civilian high-level waste.

Transuranic Waste

Before 1970, transuranic waste was not segregated from low-level waste, and a large volume of transuranic waste was buried at six federal sites. In 1970, the Atomic Energy Commission issued a directive that low-level waste containing transuranic radionuclides could no longer be disposed of by shallow land burial. Since that time, transuranic wastes have been placed in retrievable storage. The government plans to dispose of stored transuranic waste in the Waste Isolation Pilot Plant that has been built near Carlsbad, New Mexico and at which tests are being conducted to determine whether the facility can meet regulatory requirements.

Buried waste. It is very difficult to determine how much soil has been contaminated by transuranic material from broken disposal containers, liquid spills, or the past practice of pouring dilute liquid transuranic waste onto soil and relying on the soil grains to capture and immobilize the radionuclides. The contaminated sites must first be found and studied. Cleanup will be difficult because of the great volumes of soil that may need to be treated and the limits and expense of existing technology.

The transuranic waste buried before 1970 is difficult to recover since many of the older containers have broken apart and contaminated adjacent soil. General DOE policy for managing that waste will be set in the Programmatic Environmental Impact Statement now being written. Decisions about what to do at specific sites will be made through agreements with the appropriate regulatory authorities and will depend on the characteristics of the waste and the conditions at each site.

Stored waste. Stored transuranic waste is contained in a variety of package types, including metal drums and wooden and metal boxes. These packages are stored in earth-covered mounds, concrete culverts, and other types of facilities. An estimated 70 percent of the drums have been in storage for more than 10 years, and 20 to 30 percent of the drums stored in mounds contain corrosion pinholes or are badly deteriorated. Some of this waste must be repackaged before it can be shipped for disposal. Since a repository is not available for disposal, DOE is planning to expand transuranic storage capacity at some sites.

However, past records about the content of transuranic waste packages are inadequate and do not provide sufficient information for determining whether storage facilities meet current standards or for deciding how to prepare or treat the waste before disposal. Since the weapons complex waste is now subject to RCRA, the department also must determine which transuranic waste contains hazardous contaminants and then upgrade storage facilities to meet EPA standards for those facilities.

FIGURE 19. Photo of cross-section of transuranic waste drum.
Source: USDOE.

About 60,000 cubic meters of defense transuranic waste is currently stored at 10 DOE national laboratories. About 60 percent of the total volume is at Idaho National Engineering Laboratory, but this waste contains little more than 20 percent of the total curies in stored defense transuranic waste. Most of the INEL waste contains only alpha-emitting radioactive elements, so the waste package provides sufficient shielding to protect workers and the environment. A small percentage (approximately 3 percent by volume) contains sufficient penetrating radiation to require remote handling. The volume of transuranic waste stored at the Savannah River Site (6.6 percent by volume) and at Oak Ridge National Laboratory (3.3 percent by volume) is small, but it represents a high percentage of the total curies and must be handled remotely. Oak Ridge stores 34 percent of the total curies and Savannah River stores about 36 percent.

Disposal: the Waste Isolation Pilot Plant. The federal government plans to dispose of defense transuranic waste in the Waste Isolation Pilot Plant, a large mined facility in deep salt beds near Carlsbad, New Mexico. The WIPP site was originally chosen by the Energy Research and Development Administration (a DOE predecessor) in 1975 as a successor to the project abandoned at Lyons, Kansas, for disposal of defense transuranic waste. Since then, the project has gone through many metamorphoses, including changes in purpose and attempts to cancel it altogether. WIPP has now emerged in a form similar to its original one: a facility for the deep geologic disposal of transuranic waste generated in U.S. defense programs.

Throughout the years since 1975, the New Mexico state government has varied in its level of support for WIPP, but the state has been consistent in its insistence on a strong oversight role and adequate safeguards against adverse environmental, public health, or economic impacts. A lawsuit brought by New Mexico in 1981 against DOE and the Department of the Interior resulted in an agreement guaranteeing "consultation and cooperation" between DOE and the state of New Mexico regarding the health and safety aspects of the project. This agreement has been amended three times as conditions and requirements of the project have changed. New Mexico's Environmental Evaluation Group, an independent state government agency funded by DOE, monitors the development of the WIPP repository, providing scientific and technical oversight for the state.

Since its use is restricted to defense waste, the facility is not subject to licensing by the NRC, but it is subject to NEPA, RCRA, and EPA high-level and transuranic waste standards. DOE issued a Final Environmental Impact Statement for WIPP in October 1980 and a Final Supplemental Environmental Impact Statement in January 1990.

The WIPP site is owned by the federal government and has been under the control of the Bureau of Land Management of the Department of the Interior. At each step in the development of the project, DOE has needed to gain permission to use the land for specific purposes and periods of time. The process and requirements to be met before DOE could gain access to proceed with final testing and eventual operation at WIPP have been controversial. DOE was unsuccessful in gaining control of the land through administrative procedure and, as recently as 1990, Congress refused to pass a law giving DOE permission to proceed with the project. The debate concerned several issues, including compensation for the state of New Mexico and funds for constructing highways to bypass certain areas; the amount of radioactive waste that can be placed in WIPP before DOE demonstrates that the facility can meet EPA's disposal and no-migration standards for mixed transuranic waste; technical and safety issues related to experiments that DOE proposed to conduct at the site; and independent (non-DOE) regulation of the facility.

In October 1992, Congress passed the WIPP Land Withdrawal Act, resolving some of these issues and setting in place procedures for resolving others. The act assigns authority for the land to DOE and establishes a new regulatory framework for WIPP involving regulatory oversight by EPA and other federal agencies. The act also contains numerous prerequisites DOE must meet before conducting tests, transporting transuranic waste, or beginning disposal operations.

Under the WIPP Land Withdrawal Act, DOE must develop a management plan for use of the land in consultation with the Department of the Interior and the state of New Mexico; prepare a test phase plan and a retrieval plan, which must be open to the public and approved by EPA; conduct the test phase; and publish biennially a performance assessment report to be reviewed by the state, EPA, the National Academy of Sciences, and the Environmental Evaluation Group.

Under the act, EPA must issue final environmental standards for the management and disposal of spent nuclear fuel and high-level and transuranic radioactive wastes by April 30, 1993; review DOE's test and retrieval plans and determine whether they meet these environmental standards; establish criteria for compliance with the standards; and, after DOE's tests and performance assessment are complete, determine whether WIPP complies with the standards.

WIPP is 655 meters below the surface, in the salt beds of the Salado Formation. DOE has already excavated major tunnels, all four planned shafts, and seven of the 56 planned rooms. The facility is intended to house up to 6.25 million cubic feet of transuranic waste. Major technical tasks include the proper characterization of existing waste and gaining sufficient data to demonstrate compliance with EPA disposal standards and RCRA regulations. A go or no-go decision on operating WIPP is not expected before the year 2000.

Low-Level and Mixed Waste

In the 1940s and 1950s, most defense and civilian low-level waste was buried in shallow unlined trenches at sites owned and operated by the federal government. In the early 1960s, the government decided to restrict its burial sites to federal use only. Some federal government low-level waste also was packaged in 55-gallon steel drums and dumped at sea, a practice that ended in 1970. In the early years, low-level waste was disposed of with little recognition of the possible need for treatment or the presence of chemical contaminants mixed with the radioactive waste, so the low-level waste facilities accepted some waste that is now called mixed waste.

Some government burial grounds are located in humid areas where the water table is high, providing the means (groundwater) for leaching out radionuclides and moving them toward surface waters or wells. Some low-level waste containers have collapsed or broken, and waste has contaminated surrounding soil. DOE's approach to low-level and mixed low-level waste management has now changed radically, partly because of EPA's regulatory pressure to force DOE to comply with hazardous waste laws.

Low-Level Waste

DOE is following more stringent management practices at shallow land disposal facilities than in the past. In addition, the department is providing extra protection for waste disposal at two particularly humid sites by building an underground vault at the Savannah River Site, South Carolina, and an above-ground, earth-covered mound or tumulus at Oak Ridge, Tennessee. A facility is being built at Hanford, Washington, to process liquid low-level mixed waste into a cement-like material called grout. The department plans to dispose of grout in the ground at the site, although questions have been raised about the long-term stability of this waste form.

To limit the amount of low-level waste disposal capacity that is needed, DOE policy is to emphasize source reduction, volume reduction, waste segregation and improved characterization.

Although the PEIS on Environmental Restoration and Waste Management may alter planned locations for waste treatment, storage, and disposal facilities, DOE currently intends to operate low-level waste disposal facilities at sites in Hanford, Washington; the Nevada Test Site; Los Alamos National Laboratory, New Mexico; Idaho National Engineering Laboratory; Oak Ridge National Laboratory, Tennessee; and the Savannah River Site, South Carolina. At least 20 other sites that generate low-level waste will send their waste to these six sites. Those shipments raise concerns about transportation in states and communities through which the shipments may pass.

Mixed Waste

All currently operating defense low-level waste disposal sites contain what is now categorized as mixed waste. In the past, such waste was managed as low-level radioactive waste without particular regard for its chemical contaminants. Facility operators must now refit closed trenches to meet RCRA requirements and revise plans for closing the facilities. This is a very expensive, technically difficult undertaking.

To meet RCRA requirements for newly generated waste, DOE is implementing strategies that minimize the amount of mixed waste it generates, separate and keep separate any mixed waste that is generated, and store the waste pending treatment. Most defense mixed waste is being stored due to the lack of treatment and disposal facilities that meet current regulations. The technology needed to treat some types of mixed waste does not exist and may not be available for another decade.

DOE is building or has built facilities to incinerate mixed waste at Oak Ridge National Laboratory in Tennessee, Savannah River Site in South Carolina, Idaho National Engineering Laboratory, and Los Alamos National Laboratory in New Mexico. The incinerator at Oak Ridge treated 2.1 million pounds of liquid mixed waste in fiscal year 1991; the incinerator in Idaho is scheduled to be restarted in 1993. Incinerators can eliminate some chemical hazards and may leave a radioactive ash that can be disposed of as low-level, rather than mixed, waste. However, use of incineration has been controversial at federal and commercial sites.

SPECIAL PROGRAMS

DOE's Office of Environmental Restoration and Waste Management also carries out two special cleanup programs that were established before the office was created: the Formerly Utilized Sites Remedial Action Program and the Uranium Mill Tailings Remedial Action Program.

Formerly Utilized Sites Remedial Action Program

When the Manhattan Engineer District (MED), established for the Manhattan Project which developed the atomic bomb, or the Atomic Energy Commission no longer needed a facility, the AEC decommissioned it to meet the health and safety guidelines applicable at the time and then returned the facility to other uses. Those health and safety guidelines do not necessarily meet more stringent current radiological criteria for restricted and unrestricted use.

In 1974, the AEC began a survey of historical records to identify these formerly utilized sites. Most of the sites were owned by private companies and institutions. They had been used primarily for research and for processing and storing uranium and thorium ores, concentrates, and residues. In 1977, DOE formalized the program as the Formerly Utilized Sites Remedial Action Program. DOE developed a general plan to take remedial action at the sites, develop acceptable storage or disposal for waste from the sites, and certify the sites as acceptable for specific future uses.

Formerly Utilized Sites Remedial Action Program is now subject to the provisions of NEPA and CERCLA. DOE must reach agreement with EPA and state regulators on how and when the cleanup is to occur.

FUSRAP initially included 28 MED and AEC sites; Congress added five sites in 1984 and 1985, bringing the total to 33 sites in 13 states. About half of the sites are in the northeastern part of the country. Six of the sites are listed on EPA's National Priority List. Cleanup has been completed at 11 of the sites.

Most of the waste from this project will be contaminated soil and building rubble, eventually amounting to more than two million cubic yards of contaminated material classified as low-level radioactive or mixed waste. Although small amounts of transuranic nuclides may be present in some of the waste, none of the waste is expected to contain enough to be classified as transuranic waste. FUSRAP site cleanup and waste management raise contentious issues at some sites, primarily concerning whether to move the waste or to stabilize it in place and, if it is to be moved, where and how is it to go.

Uranium Mill Tailings

Uranium mill tailings comprise the largest volume of any category of radioactive waste in this country. When ore is processed to extract uranium, approximately 99 percent of the mass (containing 85 percent of the radioactivity) of the original ore remains as tailings. One principal radionuclide in the pile, thorium-230, a precursor of radon-220, has a half-life of 77,000 years. This ensures that the radioactive emissions from the tailings piles will continue for a very long time indeed. To be prudent, management of these tailings must take into account the large volumes and the persistent nature of their potential hazard.

Of all radioactive wastes, however, uranium mill tailings were for many years the most neglected. Tailings do not contain a high enough concentration of radioactive materials to fall under the legal definition of "source material," so the AEC insisted it had no jurisdiction over this part of the nuclear fuel cycle, in spite of the fact that nearly all uranium mined between 1947 and 1970 was produced for the federal government.

As a result, the piles of tailings were abandoned and left unprotected when uranium mills closed. Amounting to 27 million tons, these abandoned piles can be found in Arizona, Colorado, New Mexico, North Dakota, Oregon, Pennsylvania, South Dakota, Texas, Utah, Washington, and Wyoming. A 1976 study revealed that radium in these tailings piles had leached from two to nine feet into the subsoil and that wind had blown the tailings close to buildings and onto land where livestock and wildlife graze. The study predicted that some of the abandoned tailings piles might in the future contaminate groundwater. However, it concluded that human health impacts from most of these piles are small because, although the piles emit radon, most are located in sparsely populated areas. Exceptions included a pile originally located four miles from downtown Salt Lake City (which has been moved), and others that remain near Grand Junction, Colorado, and Durango, Colorado.

Critics fault the government most severely for allowing tailings to be used in the manufacture of building materials or as fill. In the 1960s, Grand Junction firms used tailings from a closed uranium mill to manufacture concrete, later used for local construction of buildings. For almost two decades, the 30,000 people living in these buildings were exposed to radon levels up to seven times greater than the maximum allowed for uranium miners. The federal government has provided $12 million to replace the foundations of affected homes, schools, and churches. On a smaller scale, tailings also were used in buildings in Durango, Rifle, and Riverton, Colorado; in Lowman, Idaho; in Shiprock, New Mexico; and in Salt Lake City, Utah. Many city streets and building foundations in Denver also contain tailings.

In 1978, Congress passed the Uranium Mill Tailings Radiation Control Act (UMTRA) in response to increasing public concern about possible health hazards from tailings. The act made DOE responsible for 24 inactive tailings piles left from uranium mining and milling operations conducted under contracts with the Atomic Energy Commission. The department also is charged with restoring more than 5,000 "vicinity properties," which are off-site locations contaminated by tailings material. The federal government pays 90 percent of the cleanup costs and the state pays 10 percent. Final costs probably will be much higher than the originally estimated $140 million, because some of the piles are unlikely to be stabilized and rendered innocuous in place (with a covering of loose earth or clay, for example) and will instead need to be moved and treated elsewhere.

The 1988 amendments to the act directed DOE to complete cleanup by September 1994; the department does not expect to meet that date, although it has made a start. By the end of 1991, DOE had completed work at eight tailings sites and had restored more than 4,000 vicinity properties.

DOE has completed site characterization at all UMTRA sites. Most tailings have been (or are being) stabilized at their current sites. Those located in populated areas are being moved to nearby open land. The Salt Lake City mill tailings have been moved to Clive, Utah, and the site has been cleaned; the Grand Junction tailings are being moved to Cheney Reservoir, a short distance away on a Bureau of Land Management site.

Uranium mill tailings at active mills are the responsibility of the company licensed by the NRC to operate the site. Generally they pose fewer risks than abandoned tailings because they are monitored and managed by the mill operator. Although eight mills were operating at the combined rate of 12,300 tons a day as recently as 1984, by 1992 only one mill was operating in the United States, and it is expected to close soon. Currently, most uranium used in this country is imported from Canada and Australia.

Seepage of radioactive material from tailings at active mills has occurred. Under EPA standards, corrective action programs must be developed to return the groundwater to purity levels normal for the area. Mitigative action was undertaken at 16 sites to contain the seepage and reverse the flow of groundwater. Since the uranium mining and milling industry in this country has almost ceased due to decreased demand, falling prices, and foreign competition, the process of decontamination, decommissioning, and stabilization is now under way at most of the recently active sites.

One of the worst radioactive waste spills in U.S. history occurred in July 1979 at a uranium mine and mill site on the Navajo Reservation near Churchrock, New Mexico. A muddy mixture of uranium mill tailings stored behind an earthen dam poured through a 20-foot crack in the dam and gushed into a stream. One hundred tons of mill tailings escaped during the hour it took workers to seal the crack. Traces of the spill were later found as far as 75 miles away, across the Arizona border. New Mexico health authorities ordered the owner of the mill to recover the waste and clean up any contamination.

ROLES FOR CITIZENS

"Never doubt that a small group of thoughtful, committed citizens can change the world: indeed, it is the only thing that ever has."

—MARGARET MEAD

Whether or not you live near a nuclear site, decisions about radioactive waste management may have significant effects on public health and on the economy and environment in which you and future generations will live. As a taxpayer, consumer of electric power, community resident, workers rights advocate, environmentalist, public health advocate, or parent, you have a perspective that should be heard. Your involvement can help hold government officials accountable to the people whose interests they serve. It can also improve the quality of decisions about nuclear waste, by ensuring that they incorporate important technical, social, environmental, economic, and cultural information and values.

Nuclear waste issues may be technically and bureaucratically formidable at first, but persistent effort has made citizen activism an increasingly significant factor in decisions about both civilian and defense waste. If you are interested in civilian waste storage and disposal, the most likely issues for your involvement are:

- ▼ The effort to site a high-level waste repository, now focused on Yucca Mountain, Nevada.
- ▼ The effort to site a central spent fuel storage facility, now called a monitored retrievable storage facility.
- ▼ Decisions about whether and how to expand spent fuel storage capacity at reactors.
- ▼ Decisions about increasing on-site storage or building central storage for low-level waste in states that may lose access to disposal facilities.
- ▼ Efforts to site a low-level waste facility in your state or region.

If you are interested in defense waste cleanup and disposal, the most likely issues for involvement are:

- ▼ Cleanup of sites contaminated by weapons production.
- ▼ Development of a new system for disposing of defense waste, including siting new facilities and redefining the uses of existing facilities.

This chapter suggests ways to get informed about and involved in each of the above issues. Please note that the Resources section of this *Primer* contains addresses and phone numbers for agencies and offices mentioned below.

PUBLIC INVOLVEMENT IN NUCLEAR WASTE ISSUES: GENERAL APPROACHES

Citizens can get involved by participating in agency-run programs, by using grassroots organizing methods, and by working to change existing laws.

Agency programs. Public participation activities required by law or offered by agency staff are a good starting point. Citizens should evaluate such programs critically, however, and work for improvements where they are needed. The best public involvement programs are developed cooperatively by each office or agency and members of the public, and they are geared toward the specific situation and people at hand. Effective public involvement programs share certain important characteristics, which you can look for and encourage. Effective programs are:

- Initiated during the early conceptual stage of a project;
- Integrated with the technical program and focused on real decision points;
- Interactive, emphasizing timely, two-way communication and providing means through which people can know what their impact has been;
- Open, providing clear public information in nontechnical language about the program and the decision-making process;
- Continuous over the life of the project, providing multiple formats for citizen participation;
- Broad-based, incorporating diverse elements of the community;
- Adequately funded, including provisions for funding independent technical advisors.

Some people maintain that meaningful public participation on nuclear waste issues entails not only the right to know and be heard, but also the right to act should the agency or contractor fail to carry out any phase of the substantive program adequately. Some citizens now seek to negotiate the right to monitor implementation of a disposal facility or cleanup operation through an independent technical expert, who would have the authority to halt operations should performance deviate from agreed-upon criteria.

Grassroots action. You also can use a broad range of grassroots organizing methods. Stay informed about issues and timetables. Find allies, network with existing groups, and form new groups. Understand the particular contribution you can make to solving the problem, and plan your strategy. Express your views at a time when they can make a difference, and hold officials accountable for the decisions they make. Advertise, lobby, litigate, write letters to editors, and write op-ed pieces.

Legislative methods. In the course of working with an agency on a nuclear waste program, you may find that the laws governing the agency are inadequate or misdirected. Citizen involvement in the legislative process is often the only way to achieve broad systemic changes. To accomplish such change, you can write letters, participate in meetings with elected representatives or staff, encourage media attention, vote, run for office, and serve as an elected or appointed official.

PUBLIC INVOLVEMENT IN CIVILIAN WASTE MANAGEMENT

High-Level Waste Management

Congress was explicit in the 1982 Nuclear Waste Policy Act about the importance of public participation in decision-making on high-level waste management. The act states that "State and public participation in planning and development of repositories is essential in order to promote public confidence in the safety of disposal of such waste and spent fuel." However, while the act cites many specific requirements for state and tribal participation, the opportunities for more general public participation are not so clearly spelled out. Furthermore, the 1987 amendments foreclosed some avenues by limiting site consideration to Yucca Mountain, Nevada.

To obtain general information about high-level waste issues:

- Begin with additional background reading. Ask your local library to get some of the references listed in the Resources section.
- Ask to be on the mailing list of the agency responsible for your areas of concern: the Department of Transportation, the Environmental Protection Agency, the Nuclear Regulatory Commission, the Department of Energy (DOE Headquarters in Washington, D.C., for general policy issues, transportation issues, and MRS; Nevada Field Office for repository issues).
- Join an organization that monitors and reports on developments in high-level waste management, or form your own.
- Monitor the *Federal Register* or stay in touch with organizations that do. The *Federal Register* contains notices of executive branch and regulatory agency meetings and rulemakings, proposed regulations, information on hearings and comment periods, contacts for additional information, final regulations, and effective dates. The *Register* is published daily and is available from libraries and from the U.S. Government Printing Office.
- Contact your members of Congress and ask to be kept informed about pending legislation or hearings.

To have an impact on high-level waste programs in general:

- Organize or encourage local presentations of information on such subjects as the nature of radiation; the need for nuclear-generated electricity; ethical issues related to the siting of storage and disposal facilities; the need for public participation in waste transportation issues; or the scientific and technical bases for the development of waste management systems of concern to your community. Call or write your local television and radio stations and let them know that members of the community are interested in programming on these subjects. When people become more familiar with the issues, their confidence and effectiveness increase.

- Take a position on issues of concern to you and work to build grassroots support for their adoption by decision-makers. Join a local or national organization that shares your interests—or form a group to help you with the study, analysis, and monitoring that will lead to effective participation.
- Write letters to editors, or write op-ed pieces for your local newspaper.
- Respond to opportunities to comment orally or in writing on NEPA environmental documents, proposed rules or regulations, or NRC licensing proceedings.

High-Level Waste Disposal

Public attention to high-level waste disposal at present is primarily focused on site characterization activities at Yucca Mountain. People are interested in the development of radiation standards for the site and in the eventual decision about whether Yucca Mountain is a suitable site for the nation's first high-level waste repository.

To find out more about the high-level waste repository program:

- Ask to be on the mailing list for the monthly publication, *Office of Civilian Radioactive Waste Management Bulletin*, by writing or calling the Civilian Radioactive Waste Management Information Center. Staff at the Information Center can supply material and answer questions about the Civilian Radioactive Waste Management Program. Access to INFOLINK, a computerized database of current information on the program, is also available through the Information Center.

- Visit or contact one of the DOE's Public Information Offices (in Las Vegas and several other locations). These offices answer inquiries and comments about the repository program. The offices also have libraries and exhibits that are open to the public.
- Visit the Yucca Mountain facility. DOE conducts regular public tours, which include opportunities to speak with scientists and engineers working at the site.
- Write or call the state of Nevada's Nuclear Waste Project Office. The Office responds to comments and questions relating to the repository, and, upon request, provides speakers or organizes panel discussions among representatives from various interest groups. Ask to be on the mailing list for the state of Nevada's monthly newsletter.

To have an impact on the development of a permanent high level waste repository:

- Find out what state officials are responsible for technical and policy review of DOE's proposals. Work to see that the state provides enough time and expertise for the job to be done well.
- Keep state officials informed of your views.
- Actively question and comment upon DOE activities at DOE's public briefings, which are held periodically at various communities and Native American reservations affected by the Yucca Mountain activities.
- Work to keep the issue of a permanent solution for high-level waste on the national agenda. Network with groups around the country concerned about nuclear waste issues; conduct public education campaigns to help citizens understand how this issue affects them; encourage national media attention; call or write national leaders.

Central Storage Facilities

Whether or not Yucca Mountain ultimately is chosen as the site of a permanent high-level waste repository, it will be at least 20 years before a repository is available. The federal government is now deciding whether and where to build a monitored retrievable storage facility for spent fuel.

During 1992, the Office of the United States Nuclear Waste Negotiator was seeking to negotiate an agreement with a state or Indian tribe to host an MRS. DOE was providing grants to state, tribal, and local governments to investigate whether and under what conditions they would agree to be the site of a storage facility. Local citizen participation in and oversight of the design, construction, and operation of the facility will be important bargaining issues. At the end of 1992, DOE announced that it will also seek a storage site on federally owned land. This policy may be changed by the Clinton administration.

To find out more about a monitored retrievable storage facility and about the siting process:

- ▼ Write or contact the Office of the United States Nuclear Waste Negotiator for information on the status of MRS grants and negotiations. The office also has a listing of publications by organizations and groups on all sides of the high-level waste management debate.
- ▼ Write or contact the Civilian Radioactive Waste Management Information Center for information about DOE's programs and plans for monitored retrievable storage.

To have an impact on the development of an MRS:

- ▼ If your local or tribal government is a grantee, work to ensure citizen representation on advisory or study committees. Attend meetings and keep informed of the progress of feasibility studies. Respond to requests for citizen input—let your local government know what issues concern you.

- If your local or tribal government has entered into negotiations for hosting a facility, work to insure that the agreement includes adequate requirements for public involvement and technical assistance and that it provides for adequate funding and access to information to carry out those requirements.

Storage at Reactors

A utility running out of storage for its spent fuel may decide to expand its underwater storage capacity or construct dry cask storage installations. Before doing so, the utility must apply to the NRC for a license and to the state rate-setting commission for approval of the capital expenditure. The Nuclear Regulatory Commission and state rate setting commissions have provisions for public hearings and opportunities for public input to varying degrees.

To find out more about the expansion of on-site storage capacity:

- Determine the status of on-site storage facilities at nuclear power plants of concern to you. DOE's publication entitled *Spent Fuel Storage Requirements* is a good starting point. Call or write utilities and find out their plans for dealing with any anticipated shortages in storage capacity.

To have an impact on decisions about expanding spent fuel storage capacities:

- Submit written comments for and participate in NRC or state regulatory hearings on proposed expansions of on-site storage facilities.

Low-Level Waste Management

The federal government (especially Congress), regional compacts, states, and potential host communities all have roles in low-level waste management, particularly storage and disposal.

Congress has a continuing role in low-level waste management because, as discussed in Chapter 4, Congress must approve and review regional compacts. Also, if the goals set by current law are not reached due to continued delays in facility siting and licensing, Congress could change the legislation once again.

Regional compact commissions, appointed by the governors of member states, develop the waste management systems and regulations governing waste disposal within the compact region. Those regions that need new disposal facilities must decide how and where to site them.

States chosen to build a regional facility must develop the necessary siting processes and regulations. States that do not belong to compacts must determine how the low-level waste produced within their borders will be disposed of. In either case, citizens can work to be involved in selecting the sites.

As potential sites are named, the local level of government becomes the focus of citizen attention. Siting has become a contentious issue in many communities as citizens seek to evaluate the burdens and the risks posed by such a facility, the potential for local economic growth, and the financial benefits packages offered by many low-level waste disposal contractors or host states.

To find out more about low-level waste management in your state or region:

- ▼ Find out what type and how much low-level waste is produced in your state, who is producing it, and how it is being disposed of. Publications from DOE's National Low-Level Waste Program are a good starting point.
- ▼ Find out which state agencies are responsible for policy implementation and for regulation. Ask how the state is meeting its responsibilities under the Low-Level Waste Policy Amendments Act, and what plans have been made to manage low-level waste pending compliance with its requirements. Find out whether the state has been delegated authority by the NRC to license and regulate low-

level waste disposal (i.e., whether it is an Agreement State). Ask to be on mailing lists for information or meeting notices.

- If your state belongs to a compact region, find out who represents your state on the compact commission. Determine whether decisions have been made about new facilities. Ask to be on the compact authority's mailing list for meeting notices and summaries.
- Find out what area or local organizations or individuals have been following the issue and ask them for an update.

To have an impact on low-level waste management decisions in your state or region:

- Join or form a group to monitor the development of a low-level waste management system for your state and region. Take positions on issues of concern and work to have them adopted at the local, state, or regional level. Network with other state and regional organizations to share experiences and solutions to common problems.
- Write letters to the editor of local or state newspapers. Encourage media coverage of low-level waste issues.
- Respond to opportunities for comment on proposed actions, in writing or in person at workshops or hearings.
- If your state is a member of a regional compact, determine whether the compact commission's decision-making processes are open to public scrutiny and citizen input, and whether members are sufficiently accountable for their actions. If not, organize and work for improvements.
- Review host state siting plans for public information and involvement programs. If they are inadequate, work for improvements.

To learn about and to have an impact on national low-level waste management policy:

- Let your members of Congress know you are interested in the issue. Ask to be informed about any hearings or other opportunities to be involved in national policy development.
- Make sure your opinions about proposed legislation or regulations are communicated directly to the officials responsible. Comment at public meetings and hearings.
- Put your name on the mailing lists for low-level waste issues handled by the Environmental Protection Agency, the Department of Transportation, the Nuclear Regulatory Commission, and the Department of Energy.
- Join national organizations monitoring the issue and be certain you receive their newsletters.
- Monitor and participate in NRC or Agreement State rulemaking regarding long-term interim storage of low-level waste.

PUBLIC INVOLVEMENT IN DEFENSE WASTE CLEANUP AND MANAGEMENT

The DOE's Office of Environmental Restoration and Waste Management, the office responsible for cleaning up weapons contamination and disposing of defense waste, is relatively new. Basic policies and decision-making processes are still evolving. This means that citizens who become involved in defense waste cleanup and disposal programs can have a significant impact on their development.

Public Involvement at the National Level

Advisory groups enable the involvement of external parties in DOE decision-making on a long-term, in-depth basis. At the national level, two external advisory groups, the State and Tribal Government Working Group and the Stakeholders Forum, have functioned for several years to make comments and recommendations on drafts of the Five-Year Plan, the planning document for waste management and cleanup.

A newer group, the Environmental Management Advisory Committee (EMAC) formed in 1992, provides a public forum for discussion of significant waste management and cleanup issues and makes recommendations about specific national waste management and cleanup policies and programs. EMAC's initial focus was the PEIS being developed for the entire environmental restoration and waste management program.

To get involved in national defense waste issues:

- Read the *Federal Register*, available at public libraries and in DOE reading rooms, to learn of formal comment-response processes through which you can read drafts of documents such as the Five-Year Plan or the PEIS, and send DOE comments stating your views.
- Get on the mailing list of *EM Progress*, the quarterly newsletter of DOE's Office of Environmental Restoration and Waste Management.
- If you are a high school student or the parent of one, find out if your school does or can participate in DOE's student review program, which solicits honor student input to the Five-Year Plan.
- Consult non-governmental organizations that are active on this issue.

Public Involvement in Site Cleanup

Citizen involvement in the defense waste cleanup process has focused on specific cleanup sites. DOE field offices and installations vary widely in the frequency and meaningfulness of their public participation activities.

The cleanup of some types of sites is governed by laws mandating specific public participation requirements. Refer to the previous chapter on legislation to figure out whether CERCLA, RCRA, and/or NEPA apply to the site you are interested in.

To find out more about local cleanup programs and have an impact on their development and progress:

- ▼ Check the information in the site public participation plan, which is to include information about the facility or program, a description of the public participation program, a listing of who is responsible for implementation, criteria on evaluating the effectiveness of the program, a rolling annual schedule of public participation activities, a listing of DOE officials to contact, lists of regulators and stakeholders, and addresses for information repositories and the locations of regularly scheduled meetings.
- ▼ Ask the site public participation staff whether there is a local advisory committee and, if not, how to form one. Ask whether EPA Technical Assistance Grants are available for citizen groups.
- ▼ If there is a newsletter or regular schedule of events at the site, ask to be put on the mailing list.
- ▼ Ask when the next public meeting is and participate in public meetings on site-specific planning and Five-Year Plan processes.

If your site is a CERCLA site: Ask for a copy of the Community Relations Plan for your site. Under CERCLA (Superfund) requirements, for each site, DOE staff must prepare a Community Relations Plan based on public interviews; create and maintain an information repository and publish a notice of its availability; create and maintain an administrative record and publish a notice of its availability; inform the community of the availability of technical assistance grants; prepare and publish an analysis of the proposed plan and publish a notice of its availability; conduct a public meeting and comment period; discuss changes to the proposed plan; prepare a responsiveness summary; notify the public of the final selection of a remedial action or record of decision; review and revise the Community Relations Plan; notify the public of any significant change to the final remedy; and prepare a factsheet and conduct a final public briefing.

If your site is a RCRA site: EPA guidelines require providing certain specific opportunities for public participation during the permit process under the Resource Conservation and Recovery Act. These activities include preparing or conducting a mailing list, a factsheet, a statement of basis for a decision, a public notice of permit actions and public hearing, a public comment period, a public hearing if requested, a notice of decisions, and a response to comments.

If NEPA applies to your site: National Environmental Policy Act specifies a minimum level of public involvement in the preparation of environmental decision documents, such as Environmental Assessments and Environmental Impact Statements. Requirements include involving the public in determining the scope of issues and alternatives to be considered in the analysis and in reviewing document drafts.

If your site is covered by a federal facility or other kind of formal agreement: Compliance agreements among DOE, EPA and/or a state have been negotiated at many sites and contain provisions about the process of cleaning up a site and about public information and public participation in decision-making. To find out what the agreements say, contact DOE staff, state regulators or attorneys general, or EPA regional officials.

Public Involvement In Developing a Waste Management System

DOE's Waste Management program treats, stores, and disposes of waste generated by the cleanup of contaminated defense sites, by nuclear research, by weapons production, and by the dismantlement of nuclear weapons. Public involvement in many of the waste *management* activities will occur in the context of *cleanup*, as described above, since the two kinds of activities are intertwined, and much of the advice about public involvement in cleanup also applies to public involvement in waste management.

However, there is another likely focus of public interest in defense waste management. Most of this country's facilities for producing weapons and managing the waste are aging, and, given decisions to stop producing weapons and to dismantle existing ones, many of these facilities are no longer needed. DOE is now preparing a PEIS on the reconfiguration of the weapons complex with the phasing out of some facilities and changing use of others.

The completed Waste Isolation Pilot Plant near Carlsbad, New Mexico, is intended by DOE as a disposal facility for defense-related transuranic waste now stored at DOE facilities in other states, such as Colorado and Idaho. Before it can open, however, the facility must undergo tests and be shown to meet regulatory requirements.

To stay informed about WIPP and to be involved:

- Contact New Mexico's watchdog Environmental Evaluation Group and ask for a list of publications and to be put on their mailing list.
- Contact the Board on Radioactive Waste Management of the National Academy of Sciences and ask for a list of reports from their panel on WIPP. Watch for announcements about meetings of the panel in your area and attend the public sessions.
- Contact EPA for information about the agency's plans for carrying out its regulatory responsibilities for WIPP. Attend EPA meetings and hearings and comment on issues that concern you.
- Contact DOE and ask to be put on the mailing list for information about WIPP.
- Join local and national interest groups that have followed the development of WIPP.

To have an impact on the transportation of transuranic waste to WIPP:

- ▼ Participate in meetings and workshops scheduled to be held periodically over the next several years by DOE on policies and procedures for transportation of high-level waste.
- ▼ Make the transportation issue a local concern. Work with other citizens to educate your community on proposed routes and safety standards.

CONCLUSIONS

The cleanup of contaminated sites and the siting, construction, and operation of facilities for the disposal or storage of radioactive waste may have many significant effects on communities, public and worker health, the environment, and the local and national economies. Similarly, the delay or failure to provide safe and permanent disposal sites may adversely affect the communities in which waste is now held in temporary storage—in facilities that were never intended for such long-term use. Thus, it is important not only that citizens have roles in deciding how things will be done, but also that they contribute to the process of making things happen, addressing significant problems rather than delaying action indefinitely. Deciding when action is called for is sometimes the most difficult decision. Those who have faith in the democratic process believe that the public will help ensure that the right decisions are made about how and when to act.

Some citizens will follow a decision and implementation process diligently during all its phases. Others will become involved only at major decision points or when a potential decision affects them directly. In any case, citizens can play their important roles well only if they have accurate, understandable, and timely information; sufficient time and technical support;

and an opportunity to be heard. Citizens can work to ensure that the agencies responsible provide these tools, and then they must prepare to use them wisely. By getting involved, you can help shape the ground rules—key management plans, strategies, and regulations—and thus help ensure effective and equitable policies in the future.

RESOURCES

ORGANIZATIONS

Government Agencies

► *Federal:*

Department of Energy

Office of Civilian Radioactive Waste Management, Department of Energy, 1000 Independence Avenue, SW, Washington, DC 20585. (202) 586-6842.

Civilian Radioactive Waste Management Information Center, P.O. Box 44375, Washington, DC 20026. (800) 225-6972 or (202) 488-5513.

Office of Environmental Restoration and Waste Management, 1000 Independence Avenue, SW, Washington, DC 20585. (202) 586-7709.

National Low-Level Waste Management Program, EG&G, Idaho Inc., P.O. Box 1625, Idaho Falls, ID 83415-3960. (208) 526-0483.

Uranium Mill Tailings Remedial Action (UMTRA) Project Office, 5301 Central Avenue, NE, Suite 1720, Albuquerque, NM 87108. (505) 845-4628.

Nuclear Regulatory Commission

Division of Waste Management, Office of Nuclear Material Safety & Safeguards, 1 White Flint North, Mail Stop 5E4, Washington, DC 20555. (301) 504-3432.

Uranium Recovery Field Office, 730 Simms Street, Suite 100A, Golden, CO 80401. (303) 231-5800.

Office of Technology Assessment, U.S. Congress, Washington, DC 20510. (202) 224-9241.

Office of the U.S. Nuclear Waste Negotiator, 3050 North Lakeharbor Lane, Boise, ID 83703. (208) 334-9876.

Department of the Interior

Geologic Survey Public Inquiries, 411 National Center, Reston, VA 22092. (703) 648-4460.

Earth Science Information Center, 1849 C Street, NW, Room 2650, Washington, DC 20240. (202) 208-4047.

Bureau of Land Management, 1849 C Street, NW, Washington, DC 20240. (202) 208-5717.

Department of Transportation

Research & Special Programs Administration, Office of Hazardous Materials Standards, 400 7th Street, SW, Washington, DC 20590. (202) 366-4488.

Office of Hazardous Materials Safety, 400 7th Street, SW, Washington, DC 20590. (202) 366-4350.

Environmental Protection Agency, Office of Radiation Programs, 401 M Street, SW, Mail Code ORP6601J, Washington, DC 20460. (202) 233-9320.

▶ *State:*

Nevada Nuclear Waste Project Office, Capitol Complex, Carson City, NV 89710. (702) 687-3744.

Nongovernmental Agencies

American Nuclear Society, 555 North Kensington Avenue, La-Grange Park, IL 60525. (708) 579-8265.

Battelle Memorial Institute, 505 King Avenue, Columbus, OH 43201. (614) 424-6424.

Conference of Radiation Control Program Directors, 205 Capitol Avenue, Frankfort, KY 40601.. (502) 227-4543.

Council of Energy Resource Tribes, 1999 Broadway, Suite 2600, Denver, CO 80202-5726. (303) 297-2378.

Edison Electric Institute, 701 Pennsylvania Avenue, NW, Washington, DC 20004. (202) 508-5000.

Energy Research Foundation, 537 Harden Street, Columbia, SC 29205. (803) 256-7298.

Friends of the Earth, 218 D Street, SE, Washington, DC 20003. (202) 544-2600.

Institute for Energy and Environmental Research, 6935 Laurel Avenue, Suite 204, Takoma Park, MD 20912. (301) 270-5500.

Military Production Network, 218 D Street, SE, 2nd Floor, Washington, DC 20003. (202) 544-8166.

National Conference of State Legislatures, 1560 Broadway, Suite 700, Denver, CO 80202. (303) 830-2200.

National Congress of American Indians, 900 Pennsylvania Avenue, SW, Washington, DC 20003. (202) 546-9404.

National Council on Radiation Protection and Measurements, 7910 Woodmont Avenue, Suite 800, Bethesda, MD 20814. (800) 228-2652 or (301) 657-2652.

National Governors' Association, Hall of the States, 444 North Capitol Street, Suite 267, Washington, DC 20001. (202) 624-5300.

Natural Resources Defense Council, 1350 New York Avenue, NW, Suite 300, Washington, DC 20005-4709. (202) 783-7800.

Nuclear Information and Resource Service, 1424 16th Street, NW, Suite 601, Washington, DC 20036. (202) 328-0002.

Physicians For Social Responsibility, 1000 16th Street, NW, Suite 810, Washington, DC 20036. (202) 785-3777.

Public Citizen, 215 Pennsylvania Avenue, SE, Washington, DC 20003. (202) 546-4996.

Radioactive Waste Campaign, 625 Broadway, 2nd Floor, New York, NY 10012. (212) 473-7390.

Southwest Research and Information Center, P.O. Box 4524, Albuquerque, NM 87106. (505) 262-1862.

U.S. Council for Energy Awareness (USCEA), Suite 400, 1776 I Street, NW, Washington, DC 20006-3708. (202) 293-0770.

PUBLICATIONS

Government Publications

▶ *Federal:*

Department of Energy

Annual Status Report on the Uranium Mill Tailings Remedial Action Program. DOE/EM-0001. December 1992. UMTRA Project, 5301 Central Avenue, NE, Suite 1700, Albuquerque, NM 87108. (800) 523-6495 or (505) 845-4030.

Commercial Nuclear Power 1991: Prospects for the United States and the World. Office of Coal, Nuclear, Electric and Alternate Fuels, Energy Information Administration. DOE/EIA-0438(91). 1991. 142 pages. Superintendent of Documents, U.S. Government Printing Office, Washington, DC 20402-9325. (202) 783-3238.

Directions in Low-Level Radioactive Waste Management: A Brief History of Commercial Low-Level Radioactive Waste Disposal. DOE/LLW-103. October 1990. 68 pages. National Low-Level Waste Management Program, P.O. Box 1625, Idaho Falls, ID 83415-3960. (208) 526-0676.

Environmental Restoration and Waste Management: Five Year Plan, Fiscal Years 1994-1998. FYP DOE/S-0089P. January 1993. 746 pages. National Technical Information Service, 5285 Port Royal Road, Springfield, VA 22161. (703) 487-4650.

Integrated Data Base for 1992: U.S. Spent Fuel and Radioactive Waste Inventories, Projections, and Characteristics. DOE/RW-0006, Rev. 8. October 1992. 318 pages. National Technical Information Service, 5285 Port Royal Road, Springfield, VA 22161. (703) 487-4650.

1990 Annual Report on Low-Level Radioactive Waste Management Progress, Response to Public Law 99-240. DOE/EM-0059P. September 1991. 94 pages and appendices. National Low-Level Waste Management Program, P.O. Box 1625, Idaho Falls, ID 83415-3960. (208) 526-0676.

1991 State-by-State Assessment of Low-Level Radioactive Wastes Received at Commercial Disposal Sites. DOE/LLW-152. September 1992. 135 pages. National Low-Level Waste Management Program, P.O. Box 1625, Idaho Falls, ID 83415-3960. (208) 526-0676.

Science, Society, and America's Nuclear Waste. DOE/RW-036(1-4) TG. 1992. Four-volume teacher's guide for secondary school students available from the OCRWM Information Center, Attention: Curriculum Department, P.O. Box 44375, Washington, DC 20026. (800) 225-6972 or (202) 488-5513.

Spent Fuel Storage Requirements, 1989–2040. DOE/RL-90-44. November 1990. National Technical Information Service, 5285 Port Royal Road, Springfield, VA 22161. (703) 487-4650.

Spent Nuclear Fuel Discharges from U.S. Reactors 1990. SR/CNEAF/92-01. March 1992. 192 pages. National Energy Information Administration, EI-231, 1000 Independence Avenue, SW, Washington, DC 20585. (202) 586-8800

Office of Technology Assessment:

Complex Cleanup: The Environmental Legacy of Nuclear Weapons Production. OTA-O-484. February 1991. 212 pages.

Long-Lived Legacy: Managing High-Level and Transuranic Waste at the DOE Nuclear Weapons Complex. OTA-BP-O-83. May 1991. 99 pages.

Partnerships Under Pressure: Managing Commercial Low-Level Radioactive Waste. OTA-O-426. November 1989. 164 pages.

Transportation of Hazardous Materials. OTA-SET-304. July 1986. 276 pages.

Ordering information for these publications and a catalog of other OTA publications can be obtained from Publications Orders, U.S. Congress, OTA, Washington, DC 20510-8025. (202) 224-8996.

General Accounting Office:

Nuclear Waste: Hanford Single-Shell Tank Leaks Greater Than Estimated. GAO/RCED-91-177. August 1991. 20 pages.

Nuclear Waste: Operation of Monitored Retrievable Storage Facility Is Unlikely by 1998. GAO/RCED-91-194. September 1991. 52 pages.

These reports and others on nuclear waste are available from the U.S. General Accounting Office, P.O. Box 6015, Gaithersburg, MD 20877. (202) 512-6000

Other:

Nuclear Waste: Is There a Need For Federal Interim Storage? Report of the Monitored Retrievable Storage Review Commission. November 1989. 175 pages. Superintendent of Documents, U.S. Government Printing Office, Washington, DC 20402-9325. (202) 783-3238.

Sixth Report to the U.S. Congress and the U.S. Secretary of Energy. Nuclear Waste Technical Review Board. December 1992. 125 pages. U.S. Nuclear Waste Technical Review Board, 1100 Wilson Boulevard, Suite 910, Arlington, VA 22209. (703) 235-4473.

Non-Governmental Publications

Dead Reckoning: A Critical Review of the Department of Energy's Epidemiologic Research. H. Jack Geiger, M.D., *et al.* 1992. 96 pages. $10. Physicians for Social Responsibility, 1000 16th Street, NW, Suite 810, Washington, DC 20036. (202) 785-3777.

Facing Reality: The Future of the U.S. Nuclear Weapons Complex. A Project of the Tides Foundation. May 1992. 38 pages. $2.50. Nuclear Safety Campaign, 1914 North 34th Street, Suite 407, Seattle, WA 98103. (206) 547-3175.

Health Effects of Exposure to Low-Levels of Ionizing Radiation (BEIR V Report). Commission on the Biological Effects of Ionizing Radiation, National Research Council. 1990. 421 pages. $39.95 plus $4 handling. National Academy Press, 2101 Constitution Avenue, NW, P.O. Box 285, Washington, DC 20055. (800) 624-6242 or (202) 334-3313.

High-Level Dollars, Low-Level Sense: A Critique of Present Policy for the Management of Long-Lived Radioactive Wastes and Discussion of an Alternative Approach. Arjun Makhijani and Scott Saleska. 1992. 138 pages. $12. Institute for Energy and Environmental Research, 6935 Laurel Avenue, Suite 204, Takoma Park, MD 20912. (301) 270-5500.

Nuclear Imperatives and Public Trust: Dealing with Radioactive Waste. Luther J. Carter. 1987. 473 pages. $14.95 plus $2 handling. Resources for the Future, P.O. Box 4852, Hampdon Station, Baltimore, MD 21211. (410) 516-6955.

Nuclear Waste: The Problem that Won't Go Away. Nicholas Lenssen. Worldwatch Paper 106. December 1991. 62 pages. $5. Worldwatch Institute, 1776 Massachusetts Avenue, NW, Washington, DC 20036-1904. (202) 452-1999.

Payment Due: A Reactor-by-Reactor Assessment of the Nuclear Industry's $25+ Billion Decommissioning Bill. Daniel Borson *et al.* October 1990. 88 pages. $10. Public Citizen's Critical Mass Energy Project, 215 Pennsylvania Avenue, SE, Washington, DC 20003. (202) 546-4996.

Siting Low-Level Radioactive Waste Disposal Facilities: The Public Policy Dilemma. Mary R. English. 1992. 278 pages. $49.95. Greenwood Publishing Group, 88 Post Road West, Box 5007, Westport, CT 06881. (800) 225-5800.

Understanding Radioactive Waste. Raymond L. Murray. 1989. 167 pages. $12.50. Battelle Memorial Press, 505 King Avenue, Columbus, OH 43201. (800) 451-3543 or (614) 424-6393.

Newsletters (Governmental and Nongovernmental)

EM Progress: A Report from the U.S. Department of Energy's Office of Environmental Restoration and Waste Management. DOE/EM-0084P.

Published quarterly. Free. Office of Environmental Restoration and Waste Management, EM-14, U.S. Department of Energy, Washington, DC 20585. (202) 586-3919.

High-Level Radioactive Waste Newsletter. Published bimonthly. Free. National Conference of State Legislatures, Suite 700, 1560 Broadway, Denver, CO 80202. (303) 830-2200.

Latir Nuclear Waste Calendar. Published weekly. $750 per year. Latir Energy Consultants, Route 7, Box 126-B, Old Santa Fe Trail, Santa Fe, NM 87505. (505) 984-8337.

Nevada Nuclear Waste News. Published quarterly. Free. Nevada Nuclear Waste Project Office, Capitol Complex, Carson City, NV 89710. (702) 687-3744.

Nuclear Waste News. Published weekly. $589.50 per year. Business Publishers, 951 Pershing Drive, Silver Spring, MD 20910. (800) 274-0122 or (301) 589-5103.

OCRWM Bulletin: A Report From the U.S. Department of Energy's Office of Civilian Radioactive Waste Management. DOE/RW-0372P. Published quarterly. Free. OCRWM Information Center, P.O. Box 44375, Washington, DC 20026. (800) 225-6972 or (202) 488-5513.

The Radioactive Exchange. Published 23 times a year. $549 per year. Exchange Publications, P.O. Box 5757, Washington, DC 20016. (202) 296-2814.

Radioactive Waste Campaign Report. Published quarterly. $15 per year. The Radioactive Waste Campaign, 625 Broadway, 2nd Floor, New York, NY 10012-2611. (212) 473-7390.

The Report: Defense Plant Wastes. Published biweekly. $436.50 per year. Business Publishers, 951 Pershing Drive, Silver Spring, MD 20910. (800) 274-0122 or (301) 589-6300.

Weapons Complex Monitor: Waste Management & Cleanup. Published biweekly. $995 per year. Exchange Publications, P.O. Box 5757, Washington, DC 20016. (202) 296-2814.

GLOSSARY

activation products. Atomic fragments absorbed by the steel of the reactor vessel or by minerals in the water used for cooling; give off radiation for years.

activity. The rate at which radioactive material emits radiation, stated in terms of the number of nuclear disintegrations occurring in a unit of time; the common unit of radioactivity is the curie (Ci).

agreement state. A state that has entered into an agreement with the Nuclear Regulatory Commission to assume regulatory responsibility for radioactive material under Section 274 of the Atomic Energy Act of 1954 as amended.

alpha particle. Positively charged particle emitted by certain radioactive material, made up of two neutrons and two protons. It cannot penetrate clothing or the outer layer of skin.

atom. The basic component of all matter; it is the smallest part of an element having all the chemical properties of that element. Atoms are made up of protons and neutrons (in the nucleus) and electrons.

atomic mass. The number of protons and neutrons in an atom. For instance, uranium-238 has an atomic mass of 238—92 protons and 146 neutrons.

backfill. The material used to fill in around casks after they have been placed in a repository or shallow land burial trench.

background radiation. Radiation arising from natural radioactive material always present in the environment, including solar and cosmic radiation and radioactive elements in the upper atmosphere, the ground, building materials, and the human body.

basalt. An igneous rock of volcanic origin, usually fine-grained and black or dark gray.

bedded. Layered deposit of sediment in the form of rocks, products of weathering, organic material, and precipitates.

beta particle. A negatively charged particle emitted in the radioactive decay of certain nuclides. A beta particle has mass and charge equal to that of an electron. It has a short range in air and low ability to penetrate other materials.

boiling water reactor. A light-water-cooled reactor in which the water coolant that passes through the reactor is converted to high-pressure steam that flows through the turbine.

breeder reactor. A reactor that produces more fissile material than it consumes (by a process called breeding).

canister. The outermost container into which vitrified high-level waste or spent fuel rods are to be placed. Made of stainless steel or an inert alloy.

cask. Container that provides shielding during transportation of canisters of radioactive material. Usually measures 12 feet in diameter by 22 feet long and weighs 200 tons.

chain reaction (controlled). A self-sustaining series of nuclear fissions taking place in a reactor core. Neutrons produced in one fission cause the next fission.

civilian waste. Low-level and high-level (including spent fuel) radioactive waste generated by commercial nuclear power plants, manufacturing industries, and institutions (hospitals, universities, research institutions).

cladding. Protective alloy shielding in which fissionable fuel is inserted. Cladding is relatively resistant to radiation and to the physical and chemical conditions in a reactor core. The cladding may be of stainless steel or an alloy such as zircaloy.

curie. A measure of the rate of radioactive decay; it is equivalent to the radioactivity of one gram of radium or 37 billion disintegrations per second. A nanocurie is one billionth of a curie; a picocurie is one trillionth of a curie.

daughter product. Nuclides resulting from the radioactive decay of other nuclides. A daughter product may be either stable or radioactive.

decay. Disintegration of the nucleus of an unstable nuclide by spontaneous emission of charged particles, photons, or both.

decommissioning. Preparations taken for retirement of a nuclear facility from active service, accompanied by a program to reduce or stabilize radioactive contamination.

decontamination. The removal of radioactive material from the surface of or from within another material.

defense waste. Radioactive waste resulting from weapons research and development, the operation of naval reactors, the production of weapons material, the reprocessing of defense spent fuel, and the decommissioning of nuclear-powered ships and submarines.

disposal. Permanent removal from the human environment with no provision for continuous human control and maintenance.

dome. A bed that arches up to form a rounded peak deposit, e.g., a salt dome.

dose. A quantity of radiation or energy absorbed; measured in rads.

dry cask storage. Heavily shielded, air-cooled storage casks for storing spent fuel.

exposure. A measure of ionization produced in air by X-rays or by gamma radiation. Acute exposure generally refers to a high level of exposure of short duration; chronic exposure is lower-level exposure of long duration.

fissile. Able to be split by a low-energy neutron, for example, U-235.

fission. The splitting or breaking apart of a heavy atom such as uranium. When a uranium atom is split, large amounts of energy and one or more neutrons are released.

fission products. A general term for the complex mixture of nuclides produced as a result of nuclear fission. Most, but not all, nuclides in the mixture are radioactive, and they decay, forming additional (daughter) products, with the result that the complex mixture of fission products so formed contains about 200 different isotopes of over 35 elements.

fuel cycle. The complete series of steps involved in supplying fuel for nuclear reactors. It includes mining, refining, the fabrication of fuel elements, their use in a reactor, and management of spent fuel and radioactive waste. A closed fuel cycle includes chemical reprocessing to recover the fissionable material remaining in the spent fuel; an open fuel cycle does not.

gamma radiation. Short-wavelength electromagnetic radiation emitted in the radioactive decay of certain nuclides. Gamma rays are highly penetrating.

geologic isolation. The disposal of radioactive waste deep beneath the earth's surface.

half-life. Time required for a radioactive substance to lose 50 percent of its activity by decay. The half-life of the radioisotope plutonium-239, for example, is about 24,000 years. Starting with a pound of plutonium-239, in 24,000 years there will be 1/2 pound of plutonium-239, in another 24,000 years there will be 1/4 pound, and so on. (A pound of actual material remains but it gradually becomes a stable element.)

high-level waste (HLW). Highly radioactive material, containing fission products, traces of uranium and plutonium, and other transuranic elements; it results from chemical reprocessing of spent fuel. Originally produced in liquid form, HLW must be solidified before disposal.

igneous. Formed by solidification of molten rock.

interim storage. The temporary holding of waste on or away from the generator's site when disposal space is not available. Monitoring and human control are provided, and subsequent action involving treatment, transportation, or final disposition is expected.

ion. Atomic particle, atom, or chemical radical bearing an electric charge, either negative or positive.

ion exchange. A chemical process involving the reversible interchange of various ions between a solution and a solid material. It is used to separate and purify chemicals, such as fission products or rare earths in solution. This process also takes place with many minerals found in nature and with ions in solution such as groundwater.

ionization. Removal of electrons from an atom, for example, by means of radiation, so that the atom becomes charged.

ionizing radiation. Radiation capable of removing one or more electrons from atoms it encounters, leaving positively charged particles such as alpha and beta, and nonparticulate forms such as X-rays and gamma radiation. High enough doses of ionizing radiation may cause cellular damage. Nonionizing radiation includes visible, ultraviolet, and infrared light as well as radio waves.

isotopes. Different forms of the same chemical element, which are distinguished by having different numbers of neutrons (but the same number of protons) in the nucleus of their atoms. A single element may have many isotopes. For example, uranium appears in nature in three forms: uranium-234 (142 neutrons), uranium-235 (143 neutrons), and uranium-238 (148 neutrons); each uranium isotope has 92 protons.

latent period. The period or state of seeming inactivity between the time of exposure of tissue to an acute radiation dose and the onset of the final stage of radiation sickness.

light-water reactor (LWR). A nuclear reactor cooled and moderated by water.

linear hypothesis. The assumption that any radiation causes biological damage, according to a straight-line graph of health effect versus dose.

low-level waste (LLW). Radioactive waste not classified as high-level waste, transuranic waste, spent fuel, or by-product material. Most are generally short-lived and have low radioactivity.

man-rem (also person-rem). Measures the total radiation dose received by a population. It is an average radiation dose in rems multiplied by the number of people in the population group.

mixed waste. Waste that contains both both radioactive and hazardous chemical components.

mobility. The ability of radionuclides to move through food chains in the environment.

MRS facility. A proposed government facility for the monitored retrievable storage of spent fuel from commercial power reactors. Such a facility would store the fuel temporarily, pending shipment to a repository.

neutron. Uncharged particle in a nucleus. Neutrons are used to split heavy atoms in the fission reaction.

pressured water reactor (PWR). A light-water-cooled reactor operated at high pressure without boiling.

rad (radiation absorbed dose). The amount, or dose, of ionizing radiation absorbed by any material, such as human tissue.

radiation. Particles or waves from atomic or nuclear processes (or from certain machines). Prolonged exposure to these particles and rays may be harmful.

radioactive. Of, caused by, or exhibiting radioactivity.

radioactivity. The spontaneous emission of radiation from the nucleus of an atom. Radioisotopes of elements lose particles and energy through this process of radioactive decay.

radioisotope. An unstable isotope of an element that will eventually undergo radioactive decay (i.e., disintegration).

radionuclide. A radioactive species of an atom characterized by the constitution of its nucleus; in nuclear medicine, an atomic species emitting ionizing radiation and capable of existing for a measurable time, so that it may be used to image organs and tissues.

radon. A radioactive gas produced by the decay of one of the daughters of radium. Radon is hazardous in unventilated areas because it can build up to high concentrations and, if inhaled for long periods of time, may induce lung cancer.

rem (roentgen equivalent man). Unit used in radiation protection to measure the amount of damage to human tissue from a dose of ionizing radiation.

repository. A permanent disposal facility for high-level or transuranic waste and spent fuel.

reprocessing. The process by which spent fuel is separated into waste material for disposal and into material such as uranium and plutonium to be reused.

resin. A synthetic material used for ion exchange or a high molecular-weight organic material (i.e., glue, epoxy) used to solidify liquid materials.

scintillation liquids. Organic chemical solutions that produce light when bombarded with radiation. These liquids are a major component of institutional low-level waste.

shale. Compacted clay rock.

shielding. Materials, usually concrete, water, and lead, placed around radioactive material to protect personnel against the danger of radiation.

sievert (Sv). Unit of radiation dosage equal to 100 rems.

source term. The amount and type of radioactive material released into the environment in the case of a severe nuclear accident.

spent fuel. Fuel that has been "burned" (irradiated) in a nuclear power plant's reactor to the point where it no longer contributes efficiently to the nuclear chain reaction. Spent fuel is thermally hot and highly radioactive.

storage. Operations that are designed to provide isolation and easy recovery of radioactive material, and which rely on continuous human monitoring, maintenance, and protection from human intrusion for a specified period of time.

threshold hypothesis. A radiation-dose-consequence hypothesis holding that biological radiation effects will occur only above some minimum dose.

transuranic waste (TRU). Waste material contaminated with U-233 (and its daughter products), certain isotopes of plutonium, and nuclides with atomic number greater than 92 (uranium). It is produced primarily from reprocessing spent fuel and from use of plutonium in fabrication of nuclear weapons.

tuff. A rock composed of compacted volcanic ash and dust; it is usually porous and soft.

vitrification. The conversion of high-level waste into a glassy or noncrystalline solid for subsequent disposal.

volume reduction. Various methods of waste treatment, such as evaporation for liquids or compaction for solids, aimed at reducing the volume of waste.

INDEX

☆U.S. GOVERNMENT PRINTING OFFICE : 1994 O - 365-502 : QL 2